PRACTICAL SONOCHEMISTRY
User's Guide to Applications in Chemistry and Chemical Engineering

ELLIS HORWOOD BOOKS IN ORGANIC CHEMISTRY

Series Editor: Dr J. MELLOR, University of Southampton

Belen'kii, L.I. — **Chemistry of Organosulfur Compounds: General Problems**
Beyer, H. & Walter, W. — **Advanced Organic Chemistry**
Blazej, A. & Kosik, M. — **Phytomass as a Raw Material for Chemistry**
Brown, S.S. & Kodama, Y. — **Toxicology of Metals: Clinical and Experiment Research**
Corey, E.R., Corey, J.Y. & Gasper, P.P. — **Silicon Chemistry**
Dunn, A.D. & Rudorf, W.D. — **Carbon Disulphide in Organic Chemistry**
Foá, V., Emmet, E.A., Moroni, M. & Colonksi, A. — **Occupational and Environmental Chemical Hazards: Cellular and Biochemical Indices of Monitoring Toxicity**
Fresenius, P. — **Organic Chemical Nomenclature**
Hudlicky, M. — **Chemistry of Organic Fluorine Compounds: A Laboratory Manual 2nd (Revised) Edition**
Hudlicky, M. — **Reductions in Organic Chemistry**
Keese, R. & Muller, R.K. — **Fundamentals of Preparative Organic Chemistry**
Kova, J., Krutosikova, A. & Kada, R. — **Chemistry of Heterocyclic Compounds**
Lukevics, E., Pudova, O. & Sturkovich, R. — **Molecular Structure of Organosilicon Compounds**
Maksic, Z.B. — **Modelling of Structure and Properties of Molecules**
Manitto, P. — **Biosynthesis of Natural Products**
Mason, J.M. — **Practical Sonochemistry: User's Guide to Applications in Chemistry and Chemical Engineering**
Neidleman, S.L. & Geigert, J. — **Biohalogenation: Principles, Basic Roles and Applications**
O'Hagan, D. — **The Polyketide Metabolites**
Pizey, J.S. — **Synthetic reagents:**
Volume 1: Dimethyl Formamide; Lithium Aluminium Hydride; Mercuric Oxide; Thionyl Chloride
Volume 2: *N*-Bromosuccinide; Diazomethane; Manganese Dioxide; Raney Nickel
Volume 3: Diborane; 2,3-Dichloro-5,6-Dicyanobenzoquinone DDQ; Iodine; Lead Tetra-acetate
Volume 4: Mercuric Acetate; Periodic Acid and Periodates; Sulfuryl Chloride
Volume 5: Thallium(III) Acetate and Trifluoroacetate; Ammonia; Iodine Monochloride
Lithium Aluminium Hydride (reprinted from Synthetic Reagents Volume 1)
Sakurai, H. — **Organosilicon and Bioorganosilicon Chemistry: Structure, Bonding, Reactivity and Synthetic Application**
Smith, E.H. — **Comparative Oxidation of Organic Compounds**
Sterba, V. & Panchartek, J. — **Kinetic Methods in Organic Chemistry**
Trinajstic, N., Knop, J.V., Müller, W.R. & Szymanski, K. — **Computation Chemical Graph Theory: Generation of Structures by Computer Methods**
Tundo, P. — **Continuous Flow Methods in Organic Synthesis**
Vernin, G. — **The Chemistry of Heterocyclic Flavouring and Aroma Compounds**

PRACTICAL SONOCHEMISTRY

User's Guide to Applications in Chemistry and Chemical Engineering

TIMOTHY J. MASON B.Sc., Ph.D., C.Chem., F.R.S.C.
Professor of Chemistry, Coventry Polytechnic

ELLIS HORWOOD
NEW YORK LONDON TORONTO SYDNEY TOKYO SINGAPORE

First published in 1991 by
ELLIS HORWOOD LIMITED
Market Cross House, Cooper Street,
Chichester, West Sussex, PO19 1EB, England

A division of
Simon & Schuster International Group
A Paramount Communications Company

Typeset in Times by Ellis Horwood Limited
Printed and bound in Great Britain
by Hartnolls, Bodmin, Cornwall

British Library Cataloguing-in-Publication Data

Mason, Timothy J.
Practical sonochemistry: User's guide to applications in chemistry and chemical engineering. —
(Ellis Horwood series in organic chemistry)
I. Title. II. Series
620.2078
ISBN 0–13–682642–3

Library of Congress Cataloging-in-Publication Data available

I would like to dedicate this book to my wife Cristine and my sons Matthew, Daniel and Benjamin — perhaps I will be able to spend a little more time with them now that this text is complete.

Important notice

All of the experimental procedures described in this book have been tried and tested in the laboratories of chemists experienced in sonochemical methodology. It is the view of the author that a repetition of these experiments in any laboratory should be possible without difficulty or danger; deviation from these procedures should only be attempted with caution. Like all chemical experiments, those involving sonochemistry are potentially dangerous and great care should be exercised in conducting any of them.

No responsibility is accepted on behalf of the author or the publisher for any loss or damage arising as a result of any experiments being repeated other than strictly in accordance with the procedures set out in this book.

Table of contents

Preface

This book is a companion volume to *Sonochemistry: theory, applications and uses of ultrasound in chemistry* which was co-written with my colleague J. P. Lorimer. In that text we dealt at some length with the physical effects of ultrasound in chemistry and reviewed the synthetic applications. This text is intended to serve as a laboratory guide in that it addresses the more practical aspects of the subject and theory has been kept to a minimum.

The main reasons for writing this text were twofold: firstly the desire to promote the use of sonochemistry more widely in the laboratory and secondly to show that nowadays there is a range of power ultrasonic equipment which will enable laboratory preparations to be performed on an industrial scale. A general feeling exists that it is only when industrialists adopt a new synthetic technique that it can be said to be truly 'accepted' and it is my contention that, at the time of writing, sonochemistry is precisely at this point in its evolution.

An ever-increasing number of chemists are using power ultrasound to promote synthetic reactions, yet many experimentalists experience difficulty in producing significant sonochemical effects. This can be so even when attempting to reproduce the published work of other groups. On the whole this problem appears to be the result of a lack of appreciation of the correct methodologies to adopt in order to introduce power ultrasound into a reacting medium. This in turn is almost certainly because of a lack of understanding of the principles of ultrasound — a subject which is not very common in the background of practising chemists.

In Chapter 1 you will find a general introduction to the subject of sonochemistry. This includes experimental methods which have been used for the determination of the cavitation threshold for a medium, i.e. the minimum ultrasonic power level at which cavitation can be induced and sonochemistry becomes possible. This chapter also reviews the methods available for the measurement of the ultrasonic power level entering a reaction. Chapters 2 and 3 contain practical details on the construction and method of use of ultrasonic baths and probe systems respectively while Chapter 4 explores the type of equipment which is currently available for large-scale sonochemistry. I hope that these four chapters contain the basic information necessary for the chemist to understand how to use power ultrasound in reactions and for the engineer to appreciate some of the important design implications when introducing sonochemistry or ultrasonic processing on a plant scale.

Chapter 5 consists of a collection of sonochemical experiments, some of which have been tried and tested in our own laboratories by postgraduate students working under my supervision or that of my fellow academics in the Coventry Sonochemistry Group J. P. Lorimer, J. Lindley and D. J. Walton. Other experiments have been developed by colleagues in other laboratories who have been kind enough to submit them for inclusion in this text. I thank J. Berlan, C. Degrand, J. L. Luche, and F. Patat, France; T.-S. Chou, Taiwan; T. Ando, Japan; J. V. Sinisterra Gago, Spain; O. Repic, P. Riesz and K. S. Suslick, USA; A. Henglein, Germany; P. Kruus, Canada and R. S. Davidson, UK. Lastly I must give credit to my son Daniel who proved to be so good at using my PC to draw diagrams that he ended up as the illustrator for this text.

1

An introduction to the uses of power ultrasound in chemistry

An increasing number of scientists are becoming interested in a new field of research — sonochemistry — the effect of ultrasonic sound waves on chemical reactivity. One might expect that, like most modern scientific developments, sonochemistry has drawbacks, such as the requirements for both specialized and expensive apparatus together with a considerable degree of expertise in a particular field of study. What makes sonochemistry so appealing is not simply that it is quite inexpensive to get started but that potential applications can be found which span almost the whole range of chemical technology.

1.1 THE IMPORTANCE OF SONOCHEMISTRY

It has been recognized for many years that power ultrasound has great potential for uses in a wide variety of processes in the chemical and allied industries. Reported applications include cleaning, sterilization, flotation, drying, degassing, defoaming, soldering, plastic welding, drilling, filtration, homogenization, emulsification, dissolution, deaggregation of powder, biological cell disruption, extraction, crystallization and of course as a stimulus for chemical reactions. These uses have been reviewed periodically [1–8], and for all of them one can find an impressive number of papers and patents.

During the 1960s a number of studies were devoted to biological effects [2]. These continue to be of interest but the physical and chemical effects are clearly gaining in importance as we enter the 1990s [5–7]. Over the last five years the increasing interest in the specific uses of power ultrasound to affect organic, inorganic and organometallic synthesis (sonochemistry) has resulted in the organisation of several major conferences and the establishment of special interest groups. Amongst these events was, in the UK, the first ever international meeting devoted to sonochemistry held in 1986 [9] which was followed in the next year by the founding of a Sonochemistry Group within the Royal Society of Chemistry. In Spain in 1989 part of Ultrasonics

International was devoted to Sonochemistry [10] and, later that same year in Hawaii, a session of the Pacifichem conference was devoted to sonochemistry [11]. In France the inaugural meeting of a European Sonochemical Society was held in October 1990.

The traditional role of the synthetic chemist has been to produce new compounds for academia or industry and these materials often require new methodologies for synthesis. If the target compound has already been prepared then it becomes the development chemist's aim to scale up the method or find an alternative route which is perhaps simpler, is more efficient or leads to a purer product than that used in the research laboratory. Sonochemistry offers a powerful new weapon to both types of chemist since power ultrasound can be used in its own right for synthesis or, as we will see later, it can be used to enhance existing synthetic technologies.

1.2 THE POWER OF SOUND

Sound, as a general subject for study, is traditionally found in a physics syllabus, but it is not a topic which is met in a chemistry course and so is somewhat unfamiliar to practising chemists. Sound is transmitted through a medium by inducing vibrational motion of the molecules through which it is travelling. This motion can be visualized as rather like the ripples produced when a pebble is dropped into a pool of still water. The waves move but the water molecules which constitute the wave revert to their normal positions after the wave has passed. An alternative representation is provided by the effect of a sudden twitch of the end of a horizontal stretched spring. Here the vibrational energy is transmitted through the spring as a compression wave which is seen to traverse its whole length. This is just a single compression wave and it does not equate to sound itself, which is a whole series of such compression waves separated by rarefaction (stretching) waves in between. The pitch (or note) of the sound produced by this series of waves depends upon their frequency, i.e. the number of waves which pass a fixed point in unit time. For middle C this is 256 per second. In physics sound waves are often shown as a series of vertical lines whose separation represents intensity, or as a sine wave where intensity is shown by the amplitude (Fig. 1.1).

The physical effects of sound vibrations are most easily experienced by standing in front of a loudspeaker playing music at high volume. The actual sound vibrations are transmitted through the air and are not only audible but can also be sensed by the body through the skin. The bass notes are felt through the body more easily than the high notes and this is connected with the frequency of the pressure pulse creating the sound. Low frequency sound becomes audible at around 16 Hz (1 Hz = 1 Hertz = 1 cycle per second) but as the frequency of the sound is raised (becoming more treble) it becomes more difficult for the body to respond and that sensation is lost. At even higher frequencies the ear finds it increasingly difficult to respond and eventually the human hearing threshold is reached, normally around 16–18 kHz for adults; sound beyond this limit is inaudable and is defined as ultrasound. The hearing threshold is not the same for other animal species; thus dogs respond to ultrasonic whistles (silent dog whistles) and bats use frequencies well above 50 kHz for navigation (Fig. 1.2).

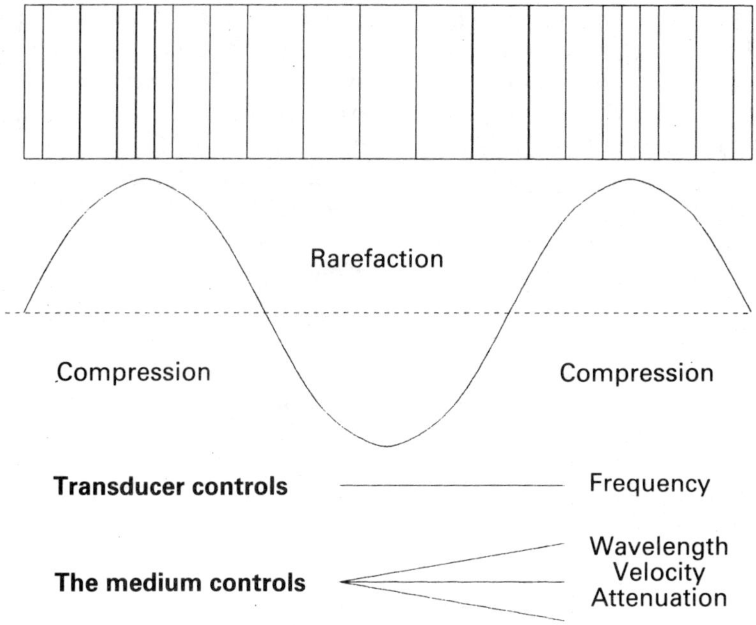

Fig. 1.1 — Sound motion in a liquid medium.

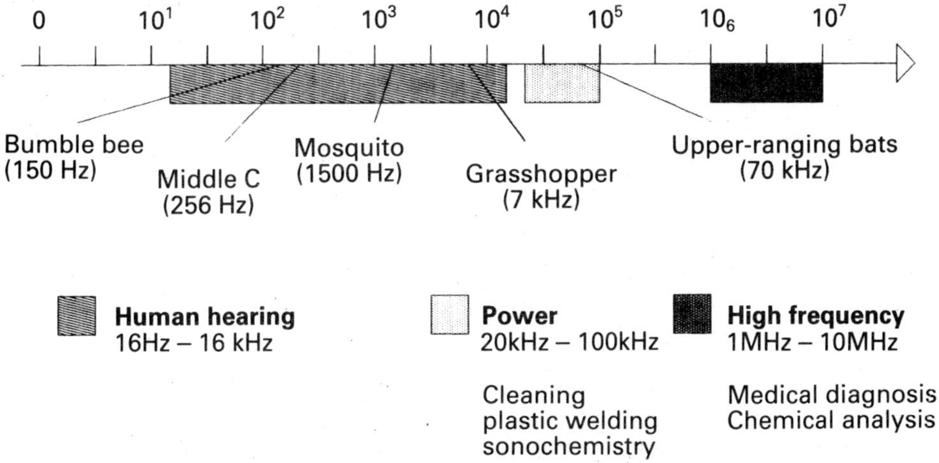

Fig. 1.2 — Sound frequencies (Hz (cycles per second)).

The broad classification of ultrasound between 18 kHz and 100 MHz has been divided into two distinct regions: power between 20 and 100 kHz and diagnostic between 1 and 10 MHz. Essentially this division arises because of the much greater sound energy which can be transmitted into a system at the lower frequencies (this induces cavitation, which is the origin of sonochemical effects). A whistle which generates a frequency of 20 kHz is inaudible to humans but perfectly audible to a dog — and produces no physical harm to either. It is, however, in the correct *frequency* range to affect chemical reactivity (power ultrasound). Yet such a whistle blown in a laboratory will not influence chemical reactions in any way. This is because such whistles do not generate enough power to influence chemical changes, i.e. the *intensity* of the sound is not sufficiently large. This aspect of ultrasound will be returned to later.

The applications of diagnostic ultrasound to chemistry are not within the scope of this text, but reference to it will be found in the companion volume to this text [4]. A summary of the uses of diagnostic ultrasound in chemistry appears in Table 1.1.

Table 1.1 — Uses of diagnostic ultrasound in chemistry

(a) The change in velocity or attenuation of sound is used:
   (i) to monitor the progress of a reaction,
   (ii) for quality control by assessing constant composition,
   (iii) for the estimation of conformational energy changes.

(b) The pulse/echo technique is used:
   (i) to measure the volumes of material in vats,
   (ii) to estimate distance to any inhomogeneous reagent interface.

(c) Changes in the absorption of sound at different frequencies can be used for the classification of biomolecule shapes.

1.3 CAVITATION — THE ORIGIN OF SONOCHEMICAL EFFECTS

It is certain that sonochemical effects cannot be the result of direct coupling of the sound field with the chemical species on the molecular level since the sound frequencies employed (20–100 kHz) are too low even for the excitation of rotational motion.

Power ultrasound produces its chemical effects through the phenomenon of cavitation. Cavitation is the production of microbubbles in a liquid when a large negative pressure is applied to it. This was first characterized at the turn of the century by Sir John Thornycroft and Sidney Barby. They were called in to investigate the poor performance of a new screw-driven destroyer — *H.M.S. Daring* and traced the problem to an incorrect setting of the propeller blades resulting in inefficient thrust. It was found that, during the rapid motion of the propeller blade through the water, the trailing edge produced sufficient negative pressure to pull the water molecules apart and create tiny microbubbles (cavities). These bubbles subsequently collapsed with the release of intense local energy. Cavitational collapse of this type near the metal surface is extremely detrimental to propeller wear since it causes erosion of the blade. Today many naval research laboratories are interested in the production of harder alloys for propellers and better design to help combat cavitational damage.

How then can power ultrasound produce cavitation? Like any sound wave ultrasound is transmitted via waves which alternately compress and stretch the molecular structure of the medium through which it passes (see Fig. 1.1). Thus the average distance between the molecules in a liquid will vary as the molecules oscillate about their mean position. If a sufficiently large negative pressure is applied to the liquid (here it is the acoustic pressure on rarefaction) that the distance between the molecules exceeds the critical molecular distance necessary to hold the liquid intact, the liquid will break down and voids will be created, i.e. cavitation bubbles will form.

Theoretical calculations indicate that extremely high energy is required to produce cavitation in water. For pure water, if the critical distance is assumed to be 10^{-8} cm then the negative pressure required is of the order of 10000 atm. If the calculation is modified so that allowance is made for the bubbles to be filled with vapour (from evaporation of the liquid), cavitation will still require negative pressures of about 1000 atm. In practice, cavitation occurs at considerably lower applied acoustic pressures owing to the presence in the liquid of weak spots which lower its tensile strength. Weak spots include the presence of gas nuclei in the form of dissolved gas, minute suspended gas bubbles or tiny suspended particles. The tiny particles (motes) are present in all liquids except those which have been subjected to ultrafiltration.

In succeeding compression cycles these cavities can collapse violently with the release of large amounts of energy in and around these microbubbles. The 'hot-spot' theory suggests that temperatures of up to 5000 K and pressures of several thousand atmospheres are produced during this collapse [12]. The alternative 'electrical' theory claims that lower temperatures and pressures are involved but electrical discharges accompany collapse [13]. Whichever theory is correct the fact remains that cavitation is the origin of sonochemistry, since the dramatic chemical changes associated with ultrasonic irradiation only occur above the cavitation threshold.

There are two forms of cavitation — stable and transient. Stable cavities are those which oscillate, often non-linearly, about some equilibrium size. Such bubbles have an existence of many cycles. Transient cavities generally exist for no more than a single acoustic cycle, during which time they expand to at least double their initial size before collapsing violently. It was once thought that the spectacular effects attributable to cavitation were entirely due to the collapse of transient cavities. It is now believed that the oscillation of 'stable' bubbles make a significant contribution to the overall sonochemical effect. In this text, however, we are interested only in the overall effects of 'cavitation' in general and so we will not concern ourselves with any distinction between types of cavitation.

1.4 PARAMETERS WHICH AFFECT CAVITATION

In spite of the great amount of published literature in the field of sonochemistry, only a relatively small number of papers report the influence of the reaction conditions on the rate and/or yields of chemical reactions. This is somewhat surprising since it is well-known that external parameters have a great influence on cavitation and, since cavitation is necessary to induce sonochemical reactions, it is important that those factors which influence cavitation are understood. Armed with this knowledge it should be easier to organize the experimental conditions such that the sonochemical effects are maximized.

1.4.1 Frequency

As the frequency of irradiation is increased so the rarefaction phase shortens and this will have two consequences:

(a) It will be necessary to increase the amplitude (power) of irradiation to maintain an equivalent amount of cavitation in the system. In other words more power is required at a higher frequency if the same cavitational effects are to be maintained, e.g. ten times more power is required to make water cavitate at 400 kHz than at 10 kHz. This is the reason why the frequencies generally chosen for cleaning and plastic welding are between 20 and 40 kHz. Consequently it is these same frequencies which have been adopted for sonochemistry.

(b) When the ultrasonic frequency is increased towards the megahertz region the production of cavitation in liquids decreases. The simplest explanation in qualitative terms is that at very high frequency, where the rarefaction (and compression) cycles are becoming extremely short, the finite time required for the rarefaction cycle becomes too short to permit the molecules to be pulled apart sufficiently to generate a bubble. It should also be recognized that transducers which operate at these high frequencies are not mechanically capable of generating very high ultrasonic power.

1.4.2 Solvent viscosity

The formation of voids or vapour-filled microbubbles (cavities) in a liquid requires that the negative pressure in the rarefaction region overcome the natural cohesive forces acting within the liquid. It follows therefore that cavitation should be more difficult to produce in viscous liquids where such forces are large.

1.4.3 Solvent surface tension

It might be expected that employing solvents with low surface tensions would lead to a reduction in the cavitation threshold. This is not a simple relationship, but certainly where aqueous solutions are involved the addition of a surfactant facilitates cavitation.

1.4.4 Solvent vapour pressure

It is more difficult to induce cavitation in a solvent of low vapour pressure, so for ease of cavitation a more volatile solvent is required. Unfortunately when a more volatile solvent is used more vapour enters the cavitation bubble during its formation and the bubble collapse is cushioned and less violent.

1.4.5 Temperature

Increasing the ambient temperature will raise the vapour pressure of the medium and so lead to easier cavitation but less violent collapse (as above). A further factor which must be considered is that at higher temperatures approaching solvent boiling point a large numbers of cavitation bubbles are generated concurrently. These will act as a barrier to sound transmission and dampen the the effective ultrasonic energy from the source which enters the liquid medium. If we sonicate a liquid at its boiling point we would therefore not expect to obtain any great sonochemical effects.

1.4.6 Bubbled gas

We have already referred to the fact that dissolved gas, or small gas bubbles in a fluid, can act as nuclei for cavitation. We will also see later that ultrasound can be used to degas a liquid. Thus at the outset of sonication cavitation will be relatively easy but as gas is removed cavitation becomes more difficult. Many groups deliberately bubble a gas through a sonochemical reaction in order to maintain uniform cavitation.

According to theory, the energy developed on collapse of these gas-filled bubbles will be greatest for gases with the largest ratio of specific heat. For this reason monoatomic gases (He, Ar, Ne) are used in preference to diatomics (N_2, air, O_2). Gases such as CO_2 are most unsuitable.

Increasing the gas content of a liquid leads not only to more facile cavitation but also to a reduction in the intensity of the shock wave released on the collapse of the bubble. If a soluble gas is used this will also provide a large number of nuclei in the solvent. The greater the solubility of the gas, the greater the amount which penetrates into the cavitation bubble, and the smaller the intensity of the shock wave created on bubble collapse. A further factor reducing the intensity of collapse may be that the more soluble the gas the more likely it is to redissolve in the medium during the compression phase of the acoustic cycle.

1.4.7 External (applied) pressure

Increasing the external pressure will mean that a greater ultrasonic energy is required to induce cavitation (greater rarefaction pressure required). More importantly, raising the external pressure will give rise to a larger intensity of cavitational collapse and consequently an enhanced sonochemical effect.

Cum *et al.* have studied the oxidation of indane to indan-1-one using potassium permanganate (Scheme 1.1) at two different frequencies (21.5 and 48 kHz) and

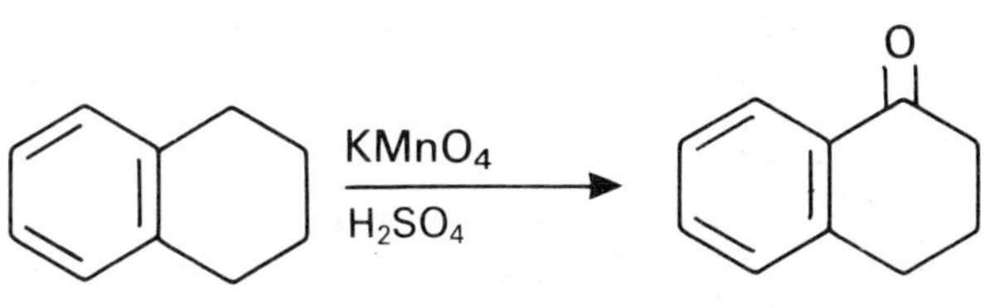

Scheme 1.1

under various external pressures [14]. For each set of experiments at a particular frequency only the applied pressure was changed while the other parameters (e.g. power, temperature, reaction volume) were kept constant. The results showed that at a specific frequency there is a particular external pressure which will provide an optimum sonochemical reaction, moreover this optimum power depends upon the frequency used.

1.4.8 Intensity

The intensity of sonication relates directly to the amplitude of vibration of the ultrasonic source. In general an increase in intensity will provide for an increase in the sonochemical effects, but ultrasonic energy input to the system cannot be increased indefinitely for three reasons:

(a) The transducer used in the sonicator will eventually break down as the increasing dimensional changes in the transducer eventually fracture the material.
(b) At high vibrational amplitude the source of ultrasound cannot maintain contact with the liquid throughout the complete cycle. Technically this is known as decoupling and results in a great loss in efficiency of transfer of power from the source to the medium.
(c) When a large amount of ultrasonic power enters a system a great number of cavitation bubbles are generated in the solution. Many of these will coalesce, forming larger, more stable bubbles. These may dampen the passage of sound energy through the liquid and also remove many of the smaller bubbles which would have collapsed to give sonochemical effects.

Luche and co-workers have extensively studied the Barbier reaction (Scheme 1.2). They have shown that the reaction rate strongly depends on both the tempera-

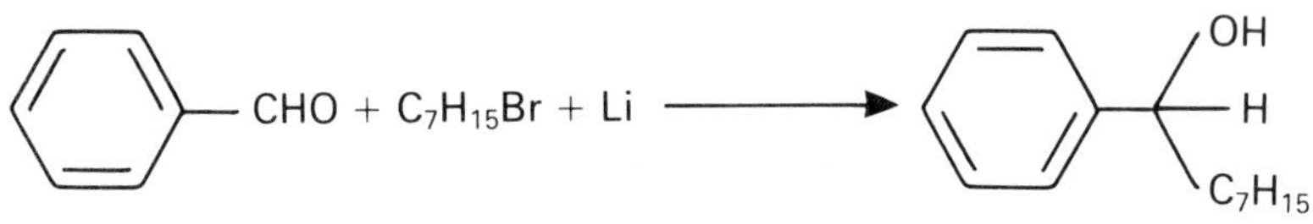

Scheme 1.2

ture and input power (intensity). Power variation was achieved by varying the applied potential at the piezoelectric transducer [15]. For both temperature and power, there is a clearly defined optimum value. A number of conclusions can be drawn from this work.

(a) A minimum intensity for sonication is required to reach the cavitation threshold; this minimum depends upon the frequency.
(b) The viscosity of a liquid medium is increased when the reaction temperature decreases. At higher viscosities cavitation is more difficult to induce (i.e. it

requires higher powers) but this is to the benefit of sonochemistry in that more violent collapse of the bubble occurs. Unfortunately at higher viscosities the number of bubbles per unit volume is reduced.

(c) When the reaction temperature is increased, the liquid viscosity decreases and the vapour pressure of the liquid increases. Under these circumstances cavitation is achieved at lower powers but the collapse will be less violent and overall the sonochemical effect will be reduced.

(d) As the intensity at the vibrating surface of the probe increases beyond an optimum value the cavitation bubble density at the resonating face, 'surface cavitation', restricts efficient transmission (coupling) of the ultrasonic energy to the bulk solution.

1.4.9 Attenuation of sound

For a variety of reasons the intensity of sound is attenuated (i.e. it decreases) as it progresses through a medium. Some of this energy is dissipated in the form of heat, although there is little appreciable heating of the bulk medium during sonication.

The extent of attenuation is inversely related to the frequency. This can be shown by using the example of sound attenuation through pure water. Sound at 118 kHz is reduced to half of its intensity after passing through 1 km of water. The distance required to achieve the same reduction of intensity for 20-kHz sound is much greater, at 30 km (this explains why submarine communications are carried out at very low frequencies). Calculations such as these demonstrate clearly that in order to achieve identical intensities at a given distance in a medium it will be necessary to use a higher initial power for the source with the higher sound frequency.

It would seem that for almost all the variable parameters which influence cavitation there will be an optimum value. This optimum can be very precise, so that the utmost care should be taken to optimize a sonochemical process.

1.5 THE TYPES OF CHEMICAL SYSTEMS AFFECTED BY POWER ULTRASOUND

Sonochemistry is mainly concerned with reactions which involve a liquid component within which cavitation can be induced, but since this covers almost all possible chemical situations, sonochemistry has very broad applications. Some typical classes of chemical reaction affected by ultrasonic irradiation are described below.

1.5.1 Homogeneous reactions — reactions involving a single liquid phase

The mechanical and chemical effects of the collapsing bubble will be felt in three distinct regions (Fig. 1.3):

(a) The inside of the bubble can be thought of as a high energy microreactor. The cavitation bubble formed in the rarefaction cycle does not enclose a vacuum — it contains vapour from the solvent (or any volatile reagent present) which enters as the bubble is being formed. On collapse, the vapour is subjected to the enormous increases in both temperature and pressure referred to above. Under

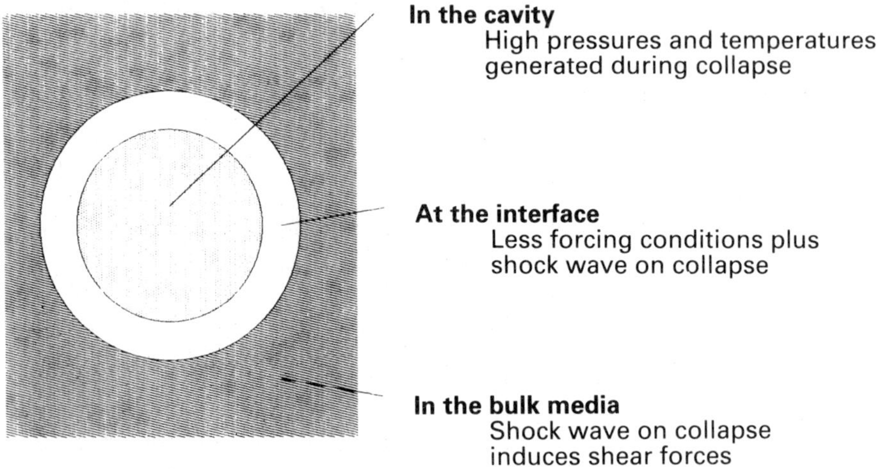

Fig. 1.3 — Cavitation effects in a homogeneous liquid.

such extreme conditions the solvent and/or reagent suffers homolytic bond breakage to generate reactive species, e.g. radicals. Thus when water is sonicated H· and HO· radicals are produced [16] which then undergo a series of reactions ultimately resulting in the production of oxygen gas and hydrogen peroxide (Scheme 1.3).

$$
\begin{aligned}
H_2O &\longrightarrow H^\bullet + {}^\bullet OH \\
H^\bullet + OH^\bullet &\longrightarrow H_2O \\
H^\bullet + H^\bullet &\longrightarrow H_2 \\
{}^\bullet OH + {}^\bullet OH &\longrightarrow H_2O_2 \\
H^\bullet + O_2 &\longrightarrow HO_2^\bullet \\
H^\bullet + HO_2^\bullet &\longrightarrow H_2O_2 \\
HO_2^\bullet + HO_2^\bullet &\longrightarrow H_2O_2 + O_2 \\
H_2O + {}^\bullet OH &\longrightarrow H_2O_2 + H^\bullet
\end{aligned}
$$

Scheme 1.3

Any species dissolved in the water is clearly going to be subject to chemical reaction with the ultrasonically produced hydrogen peroxide; thus if iodide ion is present in solution it will be oxidized and iodine will be liberated.

(b) At the interface between the bubble and the bulk liquid is a region where surface-active reagents may accumulate or where species produced in the bubble will first react with chemicals in the bulk liquid. It is clear that radicals produced in the bubble do enter the liquid phase, because when methyl methacrylate is subjected to sonication it is induced to polymerize via a free radical mechanism (Scheme 1.4) [17].

(c) In the liquid immediately surrounding the bubble an intense shock wave will be produced as liquid rapidly rushes into the volume previously occupied by the bubble. This will create enormous shear forces and these have been shown to be sufficient to fragment polymer chains dissolved in the liquid [18].

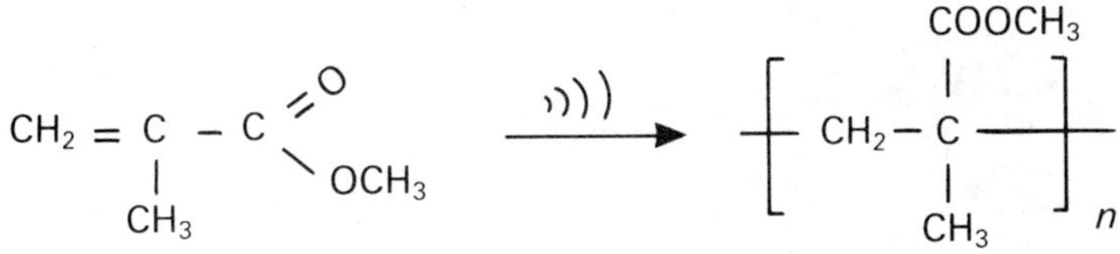

Scheme 1.4

1.5.2 Heterogeneous reactions involving a solid and a liquid

There are two types of reaction involving solid/liquid interfaces (i) in which the solid is a reagent and is consumed in the process and (ii) in which the solid — often a metal — functions as a catalyst. Two types of cavitation bubble collapse can effect the surfaces of such solids:

(a) Cavitational collapse on the surface itself which causes direct damage by the shock waves produced on implosion. These bubbles form on 'nuclei' such as surface defects, entrapped gases or impurities on the surface of the material (Fig. 1.4(a)).
(b) Cavitational collapse close to a surface which does not produce a symmetrical shock wave. Instead a jet of solvent is generated which impinges on the surface with great force — the well-known cleaning action of ultrasound (Fig. 1.4(b)).

When examined by electron microscopy surfaces of metals which have been subjected to ultrasonic irradiation reveal 'pitting'. This pitting serves both to expose new surface to the reagents and to increase the effective area available for reaction. Organometallic compounds are important intermediates for organic synthesis. They are commonly prepared by reactions between a metal and an organic compound. A general problem with such preparations is that the metal surface is easily 'poisoned' by the presence of moisture and other impurities, making reaction difficult. In most cases, therefore, the reagents used for such preparations must be pure and dry and the surfaces of the metal clean and oxide-free — as a result these syntheses require special precautions to be taken. Ultrasonic irradiation has made it possible to prepare some of these reagents even with technical-grade chemicals, conditions unheard of in classical methodology. This is of great economic importance to industry, implying that for some syntheses sonication may remove the need to employ super-pure (and thus expensive) chemicals. One example of this is the use of ultrasound for the preparation of Grignard reagents in technical-grade diethyl ether [19]. The reaction between magnesium and 2-bromobutane can be ultrasonically induced in under 5 min in this solvent, whereas conventional stirring (even with occasional crushing of the metal) takes over 3 h (Scheme 1.5).

If the solid reagent (even a metal) is in the form of a powder then sonication can cause particle rupture, with a consequent decrease in particle size and increase in surface area available for reaction. One might expect that the increase in surface area alone would be sufficient to explain any enhanced reactivity due to ultrasound — but

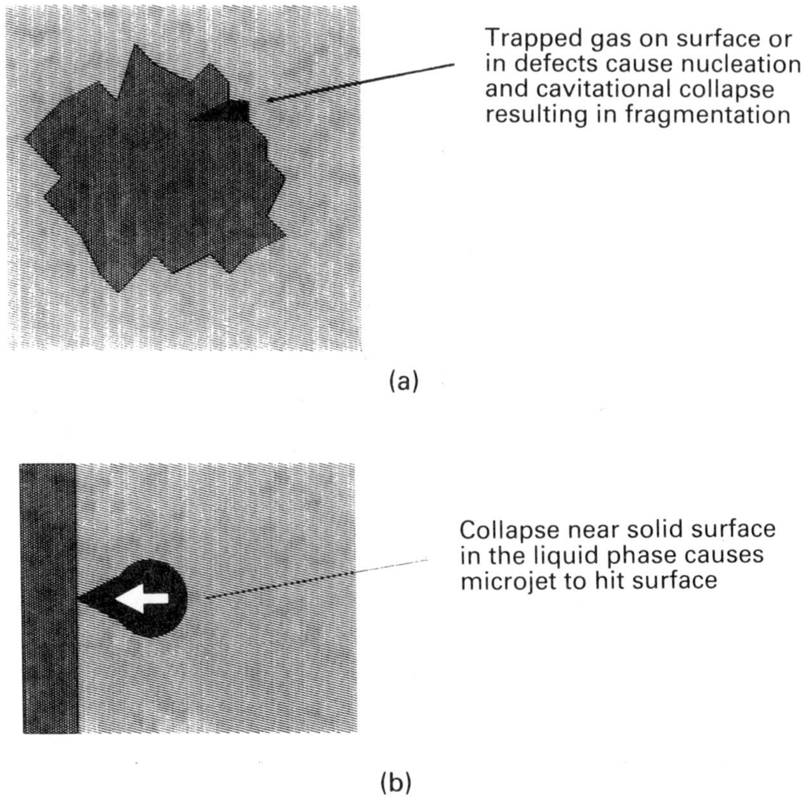

Fig. 1.4 — Cavitational effects at a solid/liquid interface: (a) powder as reagent, (b) solid catalyst surface.

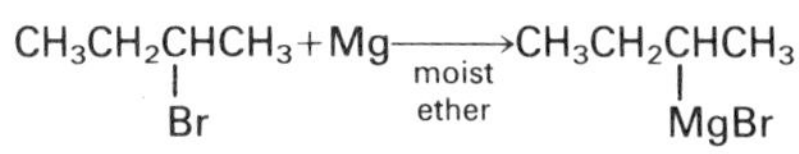

Scheme 1.5

there is more to sonochemistry than this. Consider for example the Ullmann coupling (Scheme 1.6), a reaction which was discovered many years ago and involves the prolonged heating of a halobenzene with a large excess of copper powder to produce a biphenyl. It is an expensive reaction in terms of both the heat required and the amount of copper metal which is 'wasted' in the synthesis. Under conventional conditions the reaction shown in Scheme 1.6 requires heating for 48 h and the use of a 10-fold excess of copper powder. The use of power ultrasound affords the same yield in a much shorter time (1.5 h) using only a fourfold excess of copper [20]. During

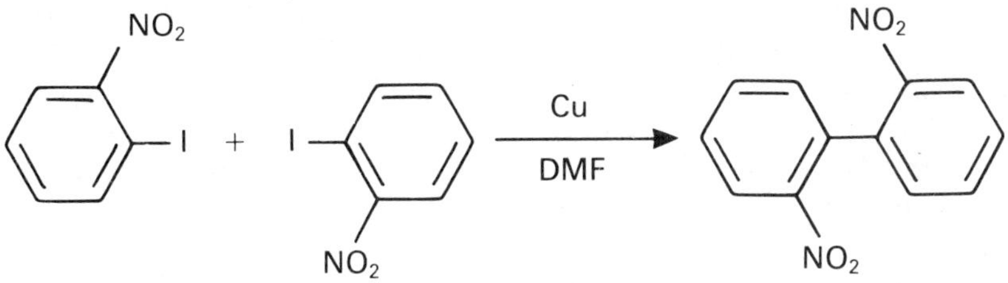

Scheme 1.6

sonication the particle size of the copper is reduced (and therefore its surface area is increased). However, when the conventional reaction is performed using copper with this reduced particle size it gives only a twofold increase compared to the large (50-fold) enhancement in reactivity produced under continuous sonication.

This study indicates another potential use for ultrasound — to 'rejuvenate' old synthetic processes which may have fallen into disuse owing to their low efficiency or expense.

1.5.3 Heterogeneous reactions involving immiscible liquids

A problem when dealing with syntheses involving immiscible liquids (e.g. aqueous/organic mixtures) is that the reagents are often dissolved in different phases. Any reaction between these species can only occur in the interfacial region between the liquids, and this is a very slow process. Sonication can be used to produce very fine emulsions from immiscible liquids. This is possible because cavitational collapse at or near the interface disrupts it and impels jets of one liquid into the other to form the emulsion (Fig. 1.5).

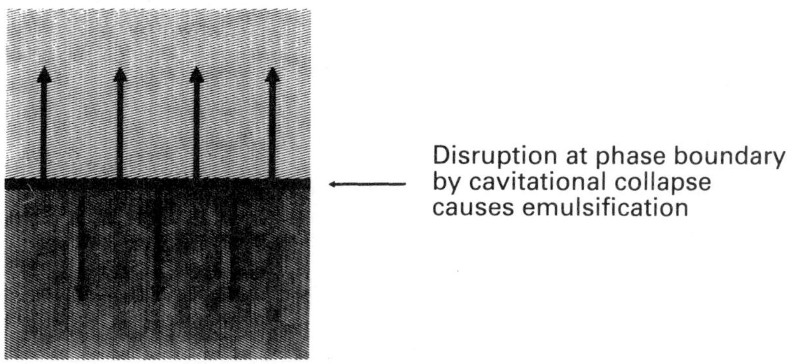

Fig. 1.5 — Cavitational effects at a liquid/liquid interface.

The synthetic chemist will induce reaction between species dissolved in different immiscible liquids (usually water and an organic solvent) by the use of a phase transfer catalyst (PTC) which will bring both reactants into the same, usually organic, phase.

There are, however, two drawbacks to the use of such catalysts in that some of the more specialized PTC reagents are expensive and all PTCs are potentially dangerous since they can, by their very nature, transfer chemicals from water into human tissue. Sonication of immiscible liquids generates extremely fine emulsions which result in very large interfacial contact areas between the liquids and a corresponding dramatic increase in the reactivity between species dissolved in the separate liquids. This effect can be used to either replace the need for a PTC or reduce dramatically the quantity required. A good example of this type of application is to be found in the hydrolysis of commercially important oils, fats and waxes in aqueous sodium hydroxide (Scheme 1.7) [21]. The fats are immiscible with the aqueous base and so there is very

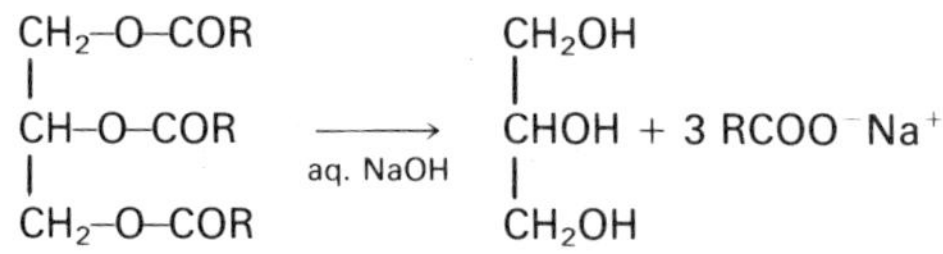

Scheme 1.7

little contact between fat and sodium hydroxide. Traditionally rapid stirring and boiling is used to increase the rate of hydrolysis but boiling tends to lead to some decomposition and the generation of coloured products. Under sonochemical conditions these hydrolyses can be carried out rapidly and at lower temperatures, leading to substantially cleaner products.

For the chemist this range of applications encompasses most of the situations encountered in synthesis. In summary we might expect to use ultrasound for a range of applications and perhaps achieve one or more of a number of beneficial effects (Table 1.2).

Table 1.2 — Beneficial effects of sonication on chemical reactivity

1. Accelerate a reaction.
2. Permit the use of less forcing conditions.
3. Make a process more economical by facilitating the use of cruder reagents.
4. Revitalize older discarded synthetic techniques by enhancing the reactivity of older reagents.
5. Reduce the number of steps required — favouring one-pot syntheses.
6. Reduce any induction period involved.
7. Enhance radical reactions.
8. Enhance catalyst efficiency.
9. Initiate stubborn reactions.

1.6 THE USES OF ULTRASOUND IN CONJUNCTION WITH OTHER TECHNIQUES

Recent decades have seen the development of a number of new technologies, each of which offers the hope of providing the chemist with innovative synthetic routes.

Some of them suffer from their own peculiar limitations. Thus photochemistry requires that the chemicals used include a chromophore, i.e. a grouping capable of being activated by light irradiation. Almost by definition, therefore, this methodology cannot work in opaque or highly coloured solutions. Power ultrasound is not attenuated by a coloured medium and does not require a specific type of functional group to be effective.

Sonochemistry certainly falls into the category of a new technology in itself but it has another major advantage for the chemist — it can be used to improve the efficiency of other methodologies. Consider the following examples of the use of ultrasound to ameliorate other processes.

1.6.1 High pressure chemistry (piezochemistry)

This requires specially reinforced, and therefore expensive, reactors. The actual reagent volumes involved are often very small (of the order of only a few millilitres for pressures above 1000 bar). In the plant the safety measures are extensive and costly. Ultrasound has been used either in place of high pressure or to reduce the overpressure required for some transformations.

$$VCl_3{\cdot}3THF \xrightarrow[\text{Na sand}]{\text{CO/THF}} V(CO)_6^-$$

Scheme 1.8

Scheme 1.8 shows the preparation of vanadium hexacarbonyl anion which requires a temperature of 200°C and 200 atm under conventional conditions. These forcing conditions can be reduced to just 10°C and only 4.4 atm when performed in the presence of power ultrasound [22]. The use of ultrasound to reduce the forcing conditions required for some transformations is not the only benefit of its use. High pressure chemistry normally only involves very small sample sizes — sonication does not suffer from this restriction.

1.6.2 Electrochemistry

This requires the presence of a conducting species to carry the current. These processes often suffer from restrictions in current flow (reduced current efficiency) during the course of the electrolysis because of electrode fouling or hindrance to ionic transport to the electrode surface. Ultrasonic irradiation has been used to great advantage to improve electrochemical processes as a result of three properties of ultrasound: it will keep the electrodes clean, degas the surface of the electrodes and increase mass transport to the surface across the electrical double layer.

Examples of this are to be found in electroplating where the benefits of ultrasonic irradiation have been known for many years [23].

1.6.3 Catalysis

Solid catalysts suffer from surface deactivation through chemical contamination (poisoning) and passivation during continuous usage. The cleaning effect of ultrasound and its surface activation are important in the enhancement of catalytic reactions. When the catalyst is in the form of a powder or an agglomerate, power ultrasound can also be used to reduce the particle size, thereby increasing the available surface area for reaction. This benefit has already been referred to above for an Ullmann reaction (Scheme 1.6) [20].

1.6.4 Supported reagents

When certain chemicals, even enzymes, are absorbed onto solid supports their reactivity is enhanced over the reagent itself. Ultrasonic irradiation provides a more efficient method of distributing and absorbing such chemicals on supports. It also enhances mass transfer to the surface. Consider the case of the substitution of bromide by cyanide (Scheme 1.9). This reaction yields no benzyl cyanide under

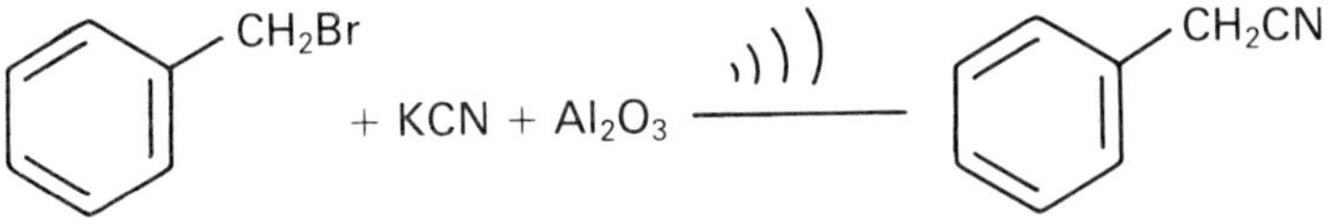

Scheme 1.9

mechanical agitation in the absence or presence of neutral alumina. However, with alumina present and using ultrasonic irradiation, the substitution product is formed in 64% yield.

A range of solid supported reactions of this type have been reviewed recently [24].

1.7 THE GENERATION OF ULTRASOUND — TRANSDUCERS

A transducer is the name for a device capable of converting one form of energy into another, a simple example being a loudspeaker which converts electrical energy to sound energy. Ultrasonic transducers are designed to convert either mechanical or electrical energy into high frequency sound and there are three main types: gas-driven, liquid-driven and electromechanical.

1.7.1 Gas-driven transducers

These are, quite simply, whistles with high frequency output. The history of the generation of ultrasound via whistles dates back 100 years to the work of F. Galton who was interested in establishing the threshold levels of human and animal hearing. Galton's whistle was constructed from a brass tube with an internal diameter of about 2 mm (Fig. 1.6) and operated by passing a jet of gas through an orifice into a resonating cavity. On moving the plunger the size of the cavity could be changed to alter the 'pitch' or frequency of the sound emitted. An adaptation of this early principle is to be found in some dog whistles which have adjustable pitch.

This type of transducer has not been found to have any significant chemical application because it is not possible to achieve a sufficiently high intensity in

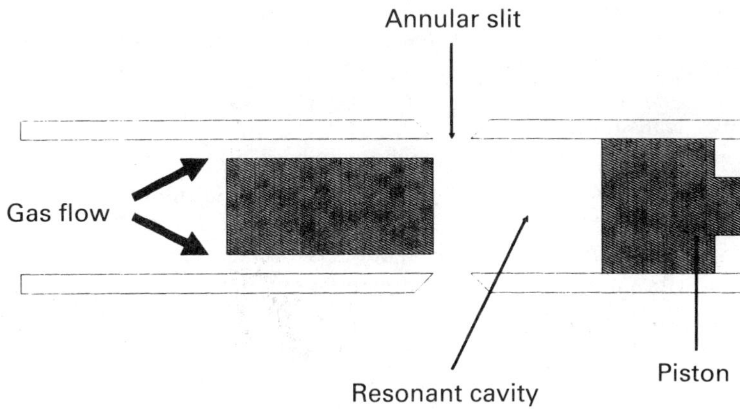

Fig. 1.6 — Schematic diagram of a Galton Whistle.

airborne ultrasound by this method. There is, however, a use for the ultrasonic whistle in the production of aerosols. When a liquid is introduced into the resonant cavity of the whistle it is atomized and the microdroplets can then be ejected as a spray (Fig. 1.7). A major application for such a device is in the formation of water mists in horticulture.

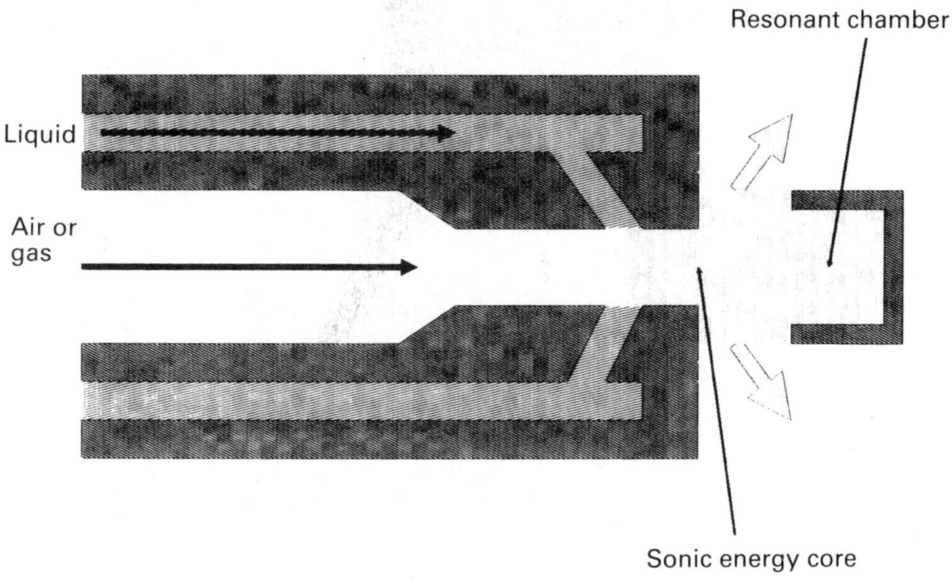

Fig. 1.7 — Compressed air spray atomizer.

Sound is attenuated rapidly on passage through air. It is, however, possible to obtain intensities of a fraction of a watt per square centimetre in gases, but the source must be very powerful — usually a sonic horn. At these intensities it is possible to use the ultrasound to break down foams, to clear fine smoke and dusts by particle agglomeration and to accelerate drying processes.

1.7.2 Liquid-driven transducers

In essence this type of transducer is a 'liquid whistle' and is effectively a propellor blade operating in inverse mode. Cavitation produced by a conventional propellor is the result of the rapid movement of the blade through water (see section 1.3 above). If, however, a metal blade is held clamped and a liquid is forced rapidly across it then the blade is caused to vibrate with a frequency dependent on the flow rate (Fig. 1.8).

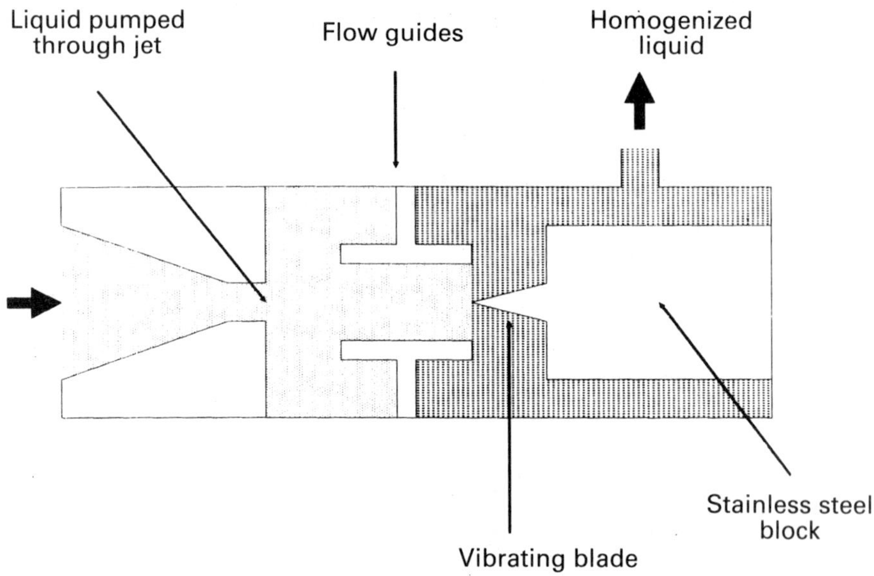

Fig. 1.8 — Liquid whistle.

When the vibrations are at ultrasonic frequencies the liquid suffers cavitation as it passes across the blade. When a mixture of immiscible liquids is forced across the blade cavitational mixing produces extremely efficient homogenization.

Equipment employing this principle has been in existence for nearly forty years. The main user has been the food industry, where the homogenization of immiscible liquids has always been important (e.g. for the production of mayonnaise or tomato ketchup). More recently chemists have turned to this simple device to provide emulsions suitable for polymerization or to enhance hydrolysis.

1.7.3 Electromechanical transducers

The two main types of electromechanical transducers are based on either the piezoelectric or the magnetostrictive effect. The most commonly used are piezoelectric transducers, generally employed to power the bath- and probe-type sonicator systems. Although more expensive than mechanical transducers, electromechanical transducers are by far the most versatile.

(i) Magnetostrictive Transducers

Magnetostriction refers to a change in the dimension of a suitable ferromagnetic material, e.g. nickel or nickel alloy, by the application of a magnetic field. Historically magnetostrictive transducers were the first to be used on an industrial scale to generate high power ultrasound. A magnetostrictive transducer is usually in the form of a rod (or bar) acting as the magnetic core within a solenoid. Applying a varying current to the coil produces a variation in the dimensions of the bar. In typical cleaning applications the nickel core is silver brazed to the stainless steel plate which couples the ultrasonic vibrations to the liquid. The unit is thus very resistant to mechanical damage since there is no obvious mode of degradation. Such transducers offer a very high driving force, are very robust and can routinely stand temperatures of up to 180°C. Traditionally they are still employed in situations where continuous high powers are required in rugged processing applications, e.g. in ore extraction and metal crystallizations [25].

The major drawback to magnetostrictive transducers is that they are not particularly efficient in terms of electrical power consumption. However, the advent of non-metallic, ceramic-based, ferrite materials (MFe_2O_4, M = divalent metal, e.g. Ni, Zn or Pb) has led to increased electrical efficiency. However, a general disadvantage of all such transducers is that their useful frequency ranges do not exceed 100 kHz.

(ii) Piezoelectric Transducers

The most common types of transducer used for both the generation and the detection of ultrasound employ piezoelectric materials. Such materials have the following two complementary properties:

(a) The direct effect — when pressure is applied across the large surfaces of the section a charge is generated on each face equal in size but of opposite sign. This polarity is reversed if tension is applied across the surfaces.
(b) The inverse effect — if a charge is applied to one face of the section and an equal but opposite charge to the other face, then the whole section of crystal will either expand or contract depending on the polarity of the applied charges.

Thus on applying rapidly reversing charges to a piezoelectric material, fluctuations in dimensions will be produced. This effect can be harnessed to transmit ultrasonic vibrations from the crystal section through whatever medium it is in contact with.

Early piezoelectric devices were based on quartz transducers, but quartz is not a particularly good material for this purpose because of its mechanical properties: it is a somewhat fragile material and is difficult to machine. Three substitutes are commonly used: barium titanate ($BaTiO_3$), lead metaniobate ($PbNb_2O_6$) and the mixed crystal lead zirconate titanate. These materials cannot be obtained as large single crystals and so, instead, they are ground with binders and sintered under pressure at above 1000°C to form a ceramic. The crystallites of the ceramic are then aligned by cooling from above the ferroelectric transition temperature in a magnetic field.

It is not possible to drive a given piece of piezoelectric material efficiently at every frequency. Optimum performance will only be obtained at the natural resonance frequency of the particular sample, and this depends upon its dimensions. In the case of quartz a transducer of thickness 2.88 mm will have a natural frequency of 1 MHz, but for 0.288 mm the frequency is 10 MHz. This is why the conventional sonochemistry equipment (baths or probes) are of fixed frequency and why reports of comparative studies at different frequencies are not common.

To reinforce the rather fragile ceramic transducers it is normal practice to clamp piezoelectric elements between metal blocks, which serve both to protect the delicate crystalline material and to prevent it from overheating by acting as a heat sink. Usually two elements are combined so that their overall mechanical motion is additive. The blocks modify, by their sheer size, the nature of the ultrasonic vibrations generated. In this way a rugged reliable transducer is obtained. In Fig. 1.9 the construction of such a transducer is shown. It is generally one half-wavelength long (although multiples of this can be used). The peak to peak amplitudes generated by such systems are normally of the order of 10–20 μm and they are electrically efficient.

Generally piezoelectric devices are not suitable for continuous usage at high temperatures. This is because the ceramic material will degrade under these conditions. For continuous use under high temperature conditions a magnetostrictive device may prove to be more reliable.

1.8 THE CAVITATION THRESHOLD

When very low power ultrasound is passed through a liquid and the power is gradually increased, a point is reached at which the intensity of sonication is sufficient to cause cavitation in the fluid. It is only at powers above the cavitation threshold that sonochemistry can occur, because only then can the great energies associated with cavitational collapse be released into the fluid. In the medical profession, where the use of ultrasonic scanning techniques is widespread, keeping scanning intensities below the cavitation threshold is of vital importance. As soon as the irradiation power used in the medical scan rises above this critical value cavitation is induced and, as a consequence, unwanted, even possibly hazardous, chemical reactions may occur in the body. Thus, for both chemical and medical reasons there is a considerable drive towards the determination of the exact point at which cavitation occurs in liquid media, particularly in aqueous systems. There are essentially three methods which can be used for the determination of the cavitation threshold, relying on either the initial detection of bubbles in the medium, the emission of light (sonoluminescence) or the initiation of a chemical reaction which occurs only in the presence of cavitation.

1.8.1 Determination by the detection of bubbles

Bubbles are extremely efficient absorbers of ultrasound and this provides the basis for a sensitive measurement of the threshold. A low intensity ultrasonic source is arranged with a receiver in a liquid such that the absorption (attenuation) of sound in that liquid can be assessed. Power ultrasound from a separate source can now be

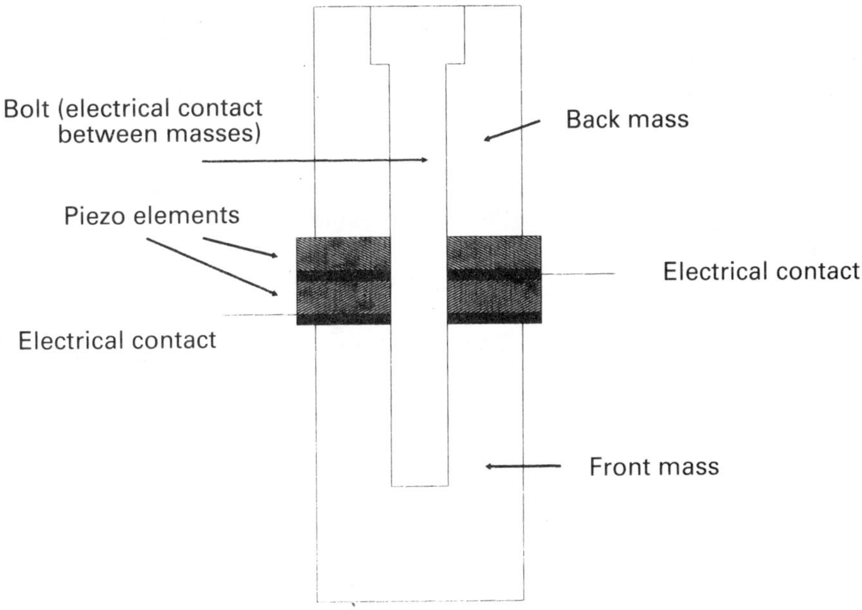

(a)

(b)

Fig. 1.9 — Sandwich transducer (a) with ceramic piezoelectric element and sandwich transducer (45 kHz) for a bath (b).

applied to the liquid and the attenuation continuously monitored as the sonication power is increased. The onset of cavitation is registered when the attenuation suddenly leaps to a higher value due to the presence of cavitation bubbles in the fluid.

1.8.2 Determination through the detection of sonoluminescence

Sonoluminescence is the emission of a weak light from cavitating fluids. This emission can be engendered by power ultrasound operating above the cavitation threshold. Sonoluminescence is outside the scope of this book, but a comprehensive review has been published recently [26]. The detection of the onset of sonoluminescence for the determination of the cavitation threshold has been successfully used by Crum *et al.* [27] who have described in detail the experimental apparatus required. This includes both a 'fluid management system' to provide ideal conditions in the liquid to be sonicated and a schematic diagram showing the circuitry used for the generation of cavitation.

1.8.3 Determination through the initiation of chemical reactions

For the practising sonochemist the above systems are perhaps too refined in that sonochemistry does not often involve ideal situations. If cavitation induces chemical reaction then this can be used as a true measure of the cavitation threshold which occurs in an actual reaction vessel, i.e. one which is later to be used for sonochemical experiments. Here we will look at three different types of reaction used as chemical probes for the cavitation threshold and sensitive dosimeters for the amount of cavitation induced by prolonged ultrasonic irradiation.

(a) Spin trapping and ESR identification of radicals

When cavitation is induced in a liquid one possible result is the production of free radicals. The hydroxy and hydrogen radicals produced by cavitation in aqueous media are short-lived species which can often be identified by spin trapping and subsequent electron spin resonance (ESR) examination of the spin adduct formed. Spin traps are usually nitroso or nitrone derivatives and the adducts are the longer-lived nitroxide radicals [28,29]. In some cases the initial short-lived radicals can be unambiguously identified. A spin trapping data base (STDB11) has been established by Buettner [30] which covers 1100 references and gives information on commonly used spin traps, the radicals trapped, the solvent, and the hyperfine coupling constants reported. For the detection of hydroxyl radicals, 5,5-dimethylpyrroline-*N*-oxide (DMPO) is particularly useful, while for hydrogen atoms, α-(4-pyridyl-1-oxide)-*N*-tert-butylnitrone (4-POBN) is preferable because of the longer lifetime of the spin adduct. For alkyl radicals, 3,5-dibromo-4-nitrosobenzene sulfonate (DBNBS), which is water-soluble and a non-volatile aromatic spin trap, is very helpful for the detection of sonochemically induced radicals. The sulfonate group ensures non-volatility, and many of the carbon-centred radical adducts show sufficiently detailed spectra to allow full identification of the trapped species. These methods have been used to study aqueous solutions of amino acids, peptides, DNA bases, sugars, nucleosides, nucleotides, surfactants, and mixtures of water and alcohols [31,32].

A description of the experimental method used can be found in Chapter 5.

(b) Using the oxidation of Fe^{2+} to Fe^{3+}

Aqueous sonochemistry can be thought of as similar in some ways to radiation chemistry. One such similarity is in the oxidation of air-saturated solution of Fe^{2+} to Fe^{3+}, a system analogous to the classical Fricke dosimeter in radiation chemistry. In 1950 Miller [33] studied this oxidation induced by ultrasound of frequency 500 kHz. The yield was found to be dependent upon the length of irradiation (up to 10 min duration), beyond which non-linearity of yield was encountered. If, however, at the end of the linear period the solution was resaturated with air and then further irradiated a linear concentration/time relationship was re-established with identical slope to that obtained previously. This observation suggests that the non-linearity is due to ultrasonically induced degassing of the solution. Clearly estimates of energy input must therefore be made based upon initial concentration changes of the Fe^{3+} ions generated. The oxidation occurs as the result of an indirect process. Cavitational collapse first generates hydroxy radicals, and these react with Fe^{2+} to yield hydroxide ion and Fe^{3+} (equation 1.1).

$$Fe^{2+} + \cdot OH \rightarrow Fe^{3+} + {}^{-}OH \qquad (1.1)$$

A typical solution would be composed of ferrous ammonium sulphate (1 mM) and sodium chloride (1 mM) dissolved in sulphuric acid (0.5 M), and the radiation dose is then calculated based on the change in absorbance of the Fe^{3+} ions at 304 nm [34] (extinction coefficient 2194 M^{-1} cm^{-1}).

(c) Using a fluorescence technique [35]

Terephthalic acid (TA) has often been used as a chemical dosimeter for ionizing radiation, and this, like the Fricke dosimeter described above, is also effectively a trap for hydroxy radicals [36]. In aqueous solution TA produces terephthalate anions which reacts with hydroxyl radicals HO· to produce highly fluorescent hydroxyterephthalate ions (HTA) (Scheme 1.10) [37]. This reaction provides a very sensitive

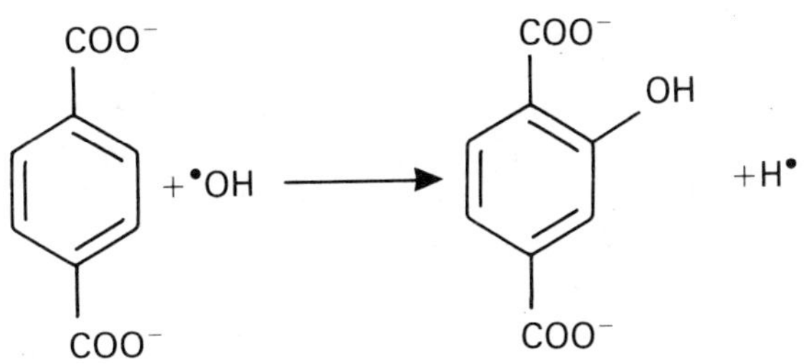

Scheme 1.10

method of measuring the hydroxyl radicals produced by the sonolysis of water and is particularly useful in determining the cavitation threshold.

TA (Prolabo) was purified and a solution prepared and stored according to the method described in the literature [38]. The following composition was employed: TA (2 mM), sodium hydroxide (5 mM) and phosphate buffer (25 mM at pH 7.4).

HTA was synthesized by alkaline fusion of bromoterephthalic acid and used to calibrate the system for fluorescence measurements at an analysing wavelength of 425 nm with an excitation wavelength of 315 nm using a Jobin Yvon spectrofluorimeter (JY3CS). Very good linearity was achieved for fluorescence versus concentration down to 3×10^{-8} M, whereas the lower limit for TA was only 2×10^{-3} M. Thus with this system it was possible to detect a 10^{-5} M fraction of TA transformed into HTA by HO· radicals. The experimental insonation equipment is shown in Fig. 1.10.

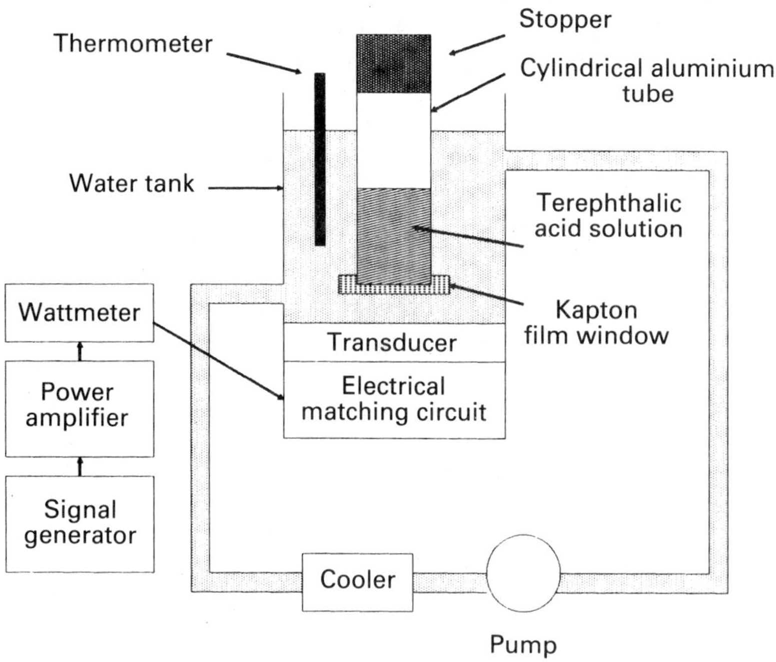

Fig. 1.10 — Equipment for terephthalic acid irradiation.

The sonicated solution was contained in a cylindrical aluminium tube coaxially aligned with the ultrasonic field. The bottom of the tube was transparent and made of Kapton film (50 μm thick). The solution volume was a 2.5 cm^3 and partly filled the 7 cm^3 container so that it freely exchanged with the air above. This exchange was enhanced by the radiation pressure, which strongly agitated the liquid–gas interface. Sonic agitation had two main consequences: first the solution was well-saturated with air and secondly there were no secondary waves in the sample since the liquid–air interface was unstable. The distance between the tube and the transducer was fixed at 8 mm. The sample holder was immersed in a water tank, with its temperature controlled by a cooling loop. Measurements of the sample temperature showed that it did not differ from the water tank temperature (20–25°C) by more than one degree.

Five different piezoceramic transducers were used at the following frequencies: 1.1, 3.6, 5.0, 6.5, and 9.5 MHz. They each generated a 15-mm-diameter beam, thus covering the surface of the sample holder, and were carefully matched to 50 Ω. The radio frequency (RF) signal was provided by a generator (Hewlett-Packard 3314A) followed by a power amplifier (Electrical Navigation Industry A150). The electrical power input was controlled by an RF Wattmeter (Rohde and Schwarz). The acoustic power spatial average temporal average (SATA) was inferred from the electrical power measurement and the efficiency, which had been previously measured in the same tank by the radiation force method using a perfect absorber [39]. The electro-acoustic efficiency ranged between 0.55 and 0.89 for the different transducers and did not vary significantly with the transferred power. The samples were sonicated for 3 min with a continuous wave. Fluorescence measurements were performed within 2 h of sonication and the samples were stored in a dark room between exposure and measurement. For each transducer several repeated experiments were undertaken with the ultrasonic intensity reduced in steps until fluorescence was no longer observed. This point was taken to be the cavitation threshold. Although the fluorescence intensity around this threshold proved to be somewhat irreproducible the actual thresholds (W cm^{-2}, SATA) were judged to be: 0.5∓0.1 (1.1 MHz); 2.0∓0.2 (3.6 MHz); 4.0∓0.3 (5.0 MHz); 7.0∓0.5 (6.5 MHz) and 13∓1 (9.5 MHz).

1.8.4 An ultrasonic exposure system for the determination of cavitation threshold

For accurate measurements of the threshold power required to generate cavitation in a fluid a well-controlled exposure environment, is essential. With the aid of such an environment the onset of cavitation can be recorded using one of the sensitive dosimeters already described. The exposure system described below [40] has two advantages. First, it generates a well-characterized acoustic beam that is reproducible. Second, it delivers both constant-wave (CW) and pulsed ultrasound, the latter being the type used in medical scanning. The range of intensities generated by this equipment simulates most therapeutic and diagnostic applications of medical ultrasound. Pulses can be generated that characterize typical diagnostic exposures, which have high peak intensity in the pulse but low overall average intensity. The onset of cavitation is signalled by the production of free radicals, and two methods for their detection are described: (a) by fluorescence monitoring and (b) by ESR spin trapping techniques [41].

The exposure tank is fabricated from 2.5-cm-thick sheets of Plexiglas and measures 18×30×22 cm (Fig. 1.11). At one end an acoustic transducer is mounted, with an absorber positioned directly opposite across the tank to minimize reflections which might result in standing waves. The absorber chosen was castor oil, since its acoustic impedance matches that of water and it has a high attenuation coefficient (in excess of 0.4 dB cm^{-1}). An acoustic window, made from a thin household plastic film (Saran Wrap, i.e. cling-film), separates the oil from the water. Better acoustic absorbing materials are now available, e.g. a silicone rubber compound filled with microballoons [42].

A constant temperature was maintained in the exposure tank to avoid thermal gradients which would disturb the acoustic beam. Bubbles were also eliminated by using degassed water (boiling distilled water under a vacuum then cooling it in a sealed vacuum bottle makes sufficiently high quality degassed water).

The reaction cell resembles an inverted T with a glass cylinder 3.3 cm in diameter and 2.5 cm in length forming the base. The stem is used for mounting and filling the

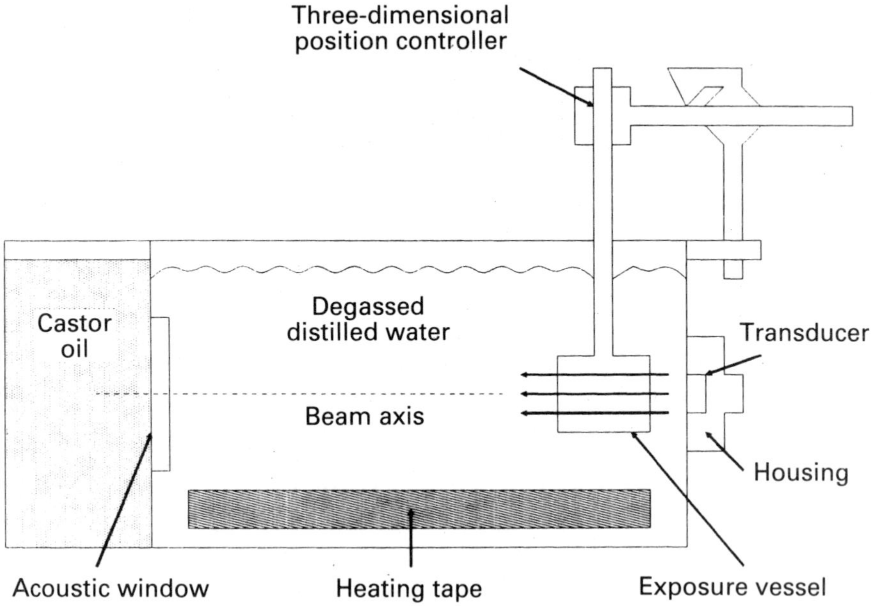

Fig. 1.11 — Ultrasonic exposure tank.

cell, which has a sample volume of 20 cm^3. Both ends of the tube are sealed with an acoustically transparent film (again Saran Wrap). Gas was bubbled through the sample prior to sonication, but this was stopped before the exposure began. The reaction cell was placed in the near field of the transducer. Although pressure variations here are more complex than in the far field [43] the positioning minimized distortion due to finite wave effects [44]. The leading edge of the tube was 6 cm from the surface of the transducer. Before each exposure the tube was centred and aligned with the axis of the acoustic beam. Data from beam plots measuring scatter in the presence and absence of the reaction cell showed that the disturbance to the acoustic beam was minimal, i.e. the acoustic beam passes smoothly and continuously through both the cell and tank.

Fig. 1.12 shows a block diagram of the experimental apparatus. An oscillator, a power amplifier, a dual directional coupler and two true RMS voltmeters generate and measure the RF power supplied to the transducer. For some applications the measurement of power is straightforward, especially when using a transducer whose electrical impedance matches that of the amplifiers; simply recording the voltage across the trandsucer is then sufficient. Pulsed exposures are more complicated, however, since they require a double-balanced mixer and a pulse generator. Using true RMS voltmeters would work for some pulse conditions but not for all. Replacing them with a transient recorder would allow accurate measurements for pulsed ultrasound with high peak amplitudes but low average exposures.

In this system an unfocussed, high powered transducer was employed with 1.0-MHz centre frequency and 2.54-cm diameter. It was acoustically matched to

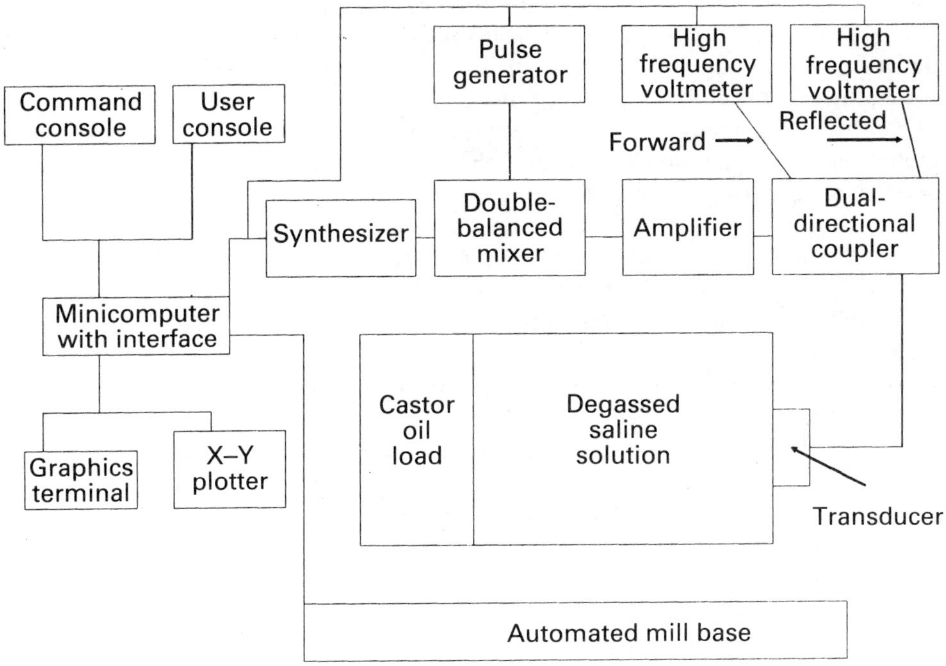

Fig. 1.12 — Block diagram of exposure tank apparatus.

water and electrically matched to 50 Ω and could continuously generate over 25 W of acoustic power. (Unfortunately, the design of transducer used here and obtained from Ultrasonic Design, Tampa, Florida, is no longer available.)

Even with a system like this, reproducible sonochemical results are not easily achieved. Controlling the exposure field does not guarantee that the sonochemical results themselves will be reproducible. Using this apparatus the acoustic field was stable and reproducible but the sonochemical results were not. Cavitation at low levels of power is an elusive phenomenon, since, apart from factors which effect the exposure field, it also depends on the state of the liquid itself [45].

1.9 ULTRASONIC POWER MEASUREMENT

An ultrasonic bath or probe system transforms electrical power into mechanical energy. This mechanical energy is then transmitted into the sonicated reaction medium. Part of it is lost to produce heat, and another part produces cavitation, but not all of the cavitational energy produces chemical and physical effects. Some energy is reflected and some is consumed in sound re-emission (harmonics and sub-harmonics). From this, it can be appreciated that the determination of an energy balance is not easy.

Any paper reporting sonochemical results would be expected to specify the make and model of the irradiation source employed, and this is normally so. Unfortunately the power input to the reacting system is sometimes reported only as the quoted maximum rating for the equipment. A 500-W unit does not mean that the instrument delivers 500 W of ultrasonic power into any system when operating at its maximum setting. In order to record the actual power input to a reaction some standard method of measurement is required. These methods may be divided into electrical measurements, which are generally used with probe systems, and chemical dosimetry, which can be applied to any sonicated system.

1.9.1 The measurement of acoustic power generated by the probe system

For probe systems acoustic power entering the treated sample is quite difficult to measure because most commercial systems provide power by demand. This means that the rated power of the generator cannot be used as an indication of acoustic power since the power transferred will depend upon (a) the type of horn attached to the transducer and (b) the area of the horn immersed in the treatment sample. There are two possible approaches to the determination of acoustic power from a probe system:

(a) The measurement of vibrational amplitude

This is the direct measurement of the amplitude at the vibrating tip of the horn, i.e. the actual mechanical motion being transmitted to the chemical reaction in which the probe is immersed. This will give a parameter that should be directly proportional to the acoustic power. The measurement of amplitude can be achieved simply by observing the ultrasonic vibration using a metallurgical microscope with a calibrated eyepiece. Since most transducers will generate amplitudes of at least 10 μm, quite accurate measurements can be made. A small spot of aluminium paint is placed on the surface of the probe and a single metallic fleck is focussed in the graticule. On turning on the power the rapidly vibrating fleck appears as a 'smear', the length of which is the amplitude of vibration. This method of measurement cannot be easily made when the probe is in use during a sonochemical experiment.

An electromechanical method is available, however, which provides continuous monitoring with a display while the probe is in use. The alternating stress in a resonating horn is at a maximum in its centre (this is the null point — the middle of a stretched elastic band). If a strain gauge is bonded across this point then the output from this will be proportional to the displacement or amplitude of vibration. This output signal can be displayed on a meter. The meter can then be calibrated by the use of a microscope as described above.

Vibrational amplitude is not an absolute measure but it does offer a very sensitive monitor for acoustic changes during a sonochemical reaction. Its major drawback is that a strain gauge is a somewhat fragile instrument; however, amplitude gives an indication of the acoustic power output rather than the electrical power into the transducer.

Using this method it is only necessary to calibrate the meter once, since any subsequent change in transducer amplitude due to loading will be accompanied by a proportional change in strain in the system. The meter will thus follow any induced changes of amplitude which occur as a result of changes in either power input or the properties of the sonicated reaction.

(b) Measurement of the real electrical power to the transducer

A wattmeter can be used to measure electrical power to the transducer. When the system is operating in a chemical reaction it is possible to derive the transmitted acoustic power by measuring electrical power to the transducer when the system is running in air (i.e. the probe tip is not immersed in a reaction mixture — a situation which is technically referred to as a system operating without a load) and then when it is loaded (i.e. dipped into the system). The transmitted electrical power is the difference between these values.

1.9.2 The measurement of ultrasonic power entering a reaction

Of the methods available to measure the amount of ultrasonic power entering a sonochemical reaction there are two which are generally used. The most common is calorimetry, which involves a measurement of the initial rate of heating produced when a system is irradiated by power ultrasound. The other method involves a chemical dosimeter, which monitors the sonochemical generation of a chemical species. The conventional system is the liberation of iodine from potassium iodide by sonochemically generated hydrogen peroxide.

(a) Calorimetry

In a pure liquid, one might assume that almost all the mechanical energy produces heat and so, via calorimetry, some estimate of the output power can be obtained. However, care should be taken in circumstances where sonochemical degradation of the liquid occurs. For example, sonicated water produces hydrogen and hydrogen peroxide, which is an endothermic reaction ($\Delta H = 383\ \mathrm{kJ\ mol^{-1}}$). If this reaction occurs in, say, a 1% yield, the absorbed energy is of the same order of magnitude as an increase in the temperature of 1 l of water by 1°C. The situation is, however, further complicated because there will be exothermic recombination reactions which follow the fragmentation. Nevertheless, this is a simple method of estimating input power.

For the system under study the temperature (T) is recorded against time (t), at 30-s intervals (or shorter), using a thermocouple placed in the reaction itself. From the T versus t data, the temperature rise ($\mathrm{d}T/\mathrm{d}t$) at zero time can be estimated either by constructing a tangent to the curve at $t=0$, or by curve-fitting the data to a polynomial in t. The ultrasonic power actually entering the system can be obtained from this by substituting the value of $\mathrm{d}T/\mathrm{d}t$ at time zero (obtained from either method) into equation (1.2)

$$\text{power} = (\mathrm{d}T/\mathrm{d}t)c_p M \tag{1.2}$$

where c_p is the heat capacity of the solvent ($\mathrm{J\ kg^{-1}\ K^{-1}}$) and M the mass of solvent used (kg).

If this power (in watts) is dissipated into the system from a probe tip with an area measured in square centimetres, then the intensity of power (in watts per square centimetre) produced by the source of ultrasound is given by equation (1.3).

$$\text{intensity} = \text{power/area} \tag{1.3}$$

(b) An iodine dosimeter for the sonication of aqueous systems

The iodine dosimetry method is based upon the fact that sonication of water produces H_2O_2 and this reacts quickly with I^- in solution to liberate molecular iodine. Thus when aliquots of standard H_2O_2 are added to KI solution, fixed quantities of iodine will be liberated and the yield of iodine can be estimated by the increase in absorbance of this species at 355 nm on a UV/visible spectrophotometer. With the aid of such a Beer–Lambert plot it is possible to calculate the concentration of H_2O_2 which gives rise to a particular absorbance of iodine and hence the yield of peroxide generated from a particular sonolysis of water.

The only difficulty with this method is that the quantity of peroxide liberated on insonation of aqueous KI is quite low [42]. Generating a Beer–Lambert calibration plot therefore requires care, with particular attention needed in ensuring that all of the glassware used is scrupulously clean. The unfortunate property of hydrogen peroxide is that it slowly decomposes on standing so that '20 volume' certainly cannot be relied upon as a standard for subsequent dilution. The analytical standard chosen for the estimation of peroxide strength is $KMnO_4$, which can be prepared by dilution of concentrated volumetric solutions and titrating in acid solution (equation 1.4).

$$2KMnO_4+5H_2O_2+3H_2SO_4=2MnSO_4+K_2SO_4+5O_2+8H_2O \qquad (1.4)$$

The standardized H_2O_2 can subsequently be diluted to the concentration required for the construction of the Beer–Lambert plot (about 1×10^{-5} mol dm^{-3}). The plot is constructed by first preparing a series of solutions from a stock of 5% acidified aqueous KI (4 g KI in 95 cm^3 distilled water plus 5 cm^3 0.5 M H_2SO_4). To each of six 10 cm^3 volumetric flasks add 5 cm^3 of the KI solution and then different quantities of the standardized H_2O_2 solution (0, 1, 2, 3, 4, 5 cm^3). Each solution is then made up to 10 cm^3 with distilled water. The solutions are shaken and left for 30 min to ensure complete reaction. The absorbance of each solution is recorded at 355 nm using 2.5% aqueous KI in the reference cell. The absorbances plotted against concentration of H_2O_2 give a very good straight-line (Beer–Lambert) calibration plot.

This calibration plot can then be used to estimate the amount of peroxide generated by various sonication systems simply by putting 5% aqueous KI in the reaction vessel and measuring the development of iodine absorbance with sonication time. In this way comparisons can be made between different pieces of sonochemical apparatus.

As we will see later, for consistent results it is important to carry out insonations with the various components of the reactor in the same configuration. In addition it is helpful if the same flask is used throughout these measurements — particularly for reactions involving ultrasonic baths — since this will help to ensure that the same amount of energy is transferred into the reaction through the glass base of the flask.

1.10 TYPES OF SONOCHEMISTRY EQUIPMENT

In subsequent chapters the various types of equipment which can be used for sonochemistry will be explored in depth. In this brief section we will simply identify the types of instrumentation currently available for sonochemistry, summarized in Table 1.3.

Table 1.3 — Types of sonicator system

1	Cleaning bath
2	Submersible transducer
3	Cup-horn
4	Probe system
5	Liquid whistle
6	Flow cell on probe
7	Tube reactor (polygon or pipe)
8	Push–pull
9	RUM

There are essentially four types of sonicator system available to the chemist and of these only the bath and probe have received any extensive use in chemistry. Both rely upon the electromagnetic (generally piezoelectric) transducer as a source of power ultrasound and suffer from the disadvantage that they operate at a fixed frequency which depends upon the particular transducer employed. For most commercial probe systems this frequency is 20 kHz; although it is possible to drive them at their overtones (i.e. 40 or 80 kHz), the power available at these frequencies is very much reduced. To operate successfully at different frequencies it is best to purchase either separate amplifier/horn systems tailored to individual requirements or a package which includes a multiple-frequency amplifier and a set of different horn systems.

1.10.1 The cleaning bath (Chapter 2)

The orthodox design involves transducers bonded to the outside of a stainless steel tank (normally on the base). The vibrations from the tank base then act as the source of ultrasound. A heater is often incorporated on larger baths, which sometimes also include a power control.

Other designs which can also be described as 'baths' include:

(a) a tank whose base has a transducer element protruding directly into the bath liquid;
(b) any tank equipped with a sealed transducer device designed to be dipped into the bath liquid (submersible transducer);
(c) the cup-horn device, which is an attachment for a probe system — essentially a small-scale version of (a).

1.10.2 The probe system (Chapter 3)

This is the best source of high power ultrasound for the laboratory. The ultrasonic energy is conducted from the transducer element directly into a reaction via a sonic horn normally made of titanium alloy. The ultrasonic energy is amplified by the horn and is produced by the vibrations of the horn tip. Probe systems are generally available only at 20 kHz frequency but some newer systems offer a choice of frequencies.

1.10.3 Flow processors and scale-up equipment (Chapter 4)

Laboratory-scale processors offer the possibility of processing quite large volumes by sonication of part of the reacting mixture as it flows through a loop system linked to a reservoir containing the reaction mixture. Several different types of flow reactor are commercially available for laboratory use:

(a) The liquid whistle homogenizer.
(b) A flow cell attachment for the probe system.
(c) A tube with transducers bonded around the outside so that the metal which forms the tube becomes the ultrasonically vibrating source.
(d) A 'push–pull' bar consisting of a cylindrical titanium rod with a transducer at each end. The action of the transducers causes rapid changes in the rod diameter, thus producing vibrations. As the centre of a flow tube this is in the form of a flow system.

Large-scale sonochemical equipment for industry can be simply a large cleaning bath for batch treatment. An approach which is receiving more attention, however, is the flow system. This can be in many forms:

(e) Tube reactors as described above but with larger diameter and more powerful transducers, or a large push–pull system.
(f) The reverberatory ultrasonic mixer (RUM) which involves two ultrasonically agitated plates forming a sandwich across a flowing system.

1.11 THE SAFETY OF SONOCHEMICAL EQUIPMENT

There is perhaps a temptation to think that if you cannot hear a noise it cannot be harmful. This is certainly true of sound in the audible range because in that situation zero ear response means very low intensity indeed. With ultrasound, however, much more care must be taken. It is important to note that most ultrasonic devices, when irradiating a liquid medium, do give rise to some audible sound. This is because the sound spectrum emitted from a sonicated reaction is not composed entirely of the single transducer resonance but includes some sub-harmonics which give rise to very audible sound levels.

Ultrasound is attenuated rapidly in air, but sources of ultrasound such as probes start at such a high intensity that there is a distinct possibility that long-term exposure without ear protection might result in damage. Consideration must therefore be given to the health and safety aspects associated with the use of ultrasound.

Permitted exposure limits do vary from country, to country but generally they lie between 85 and 90 dBA over an 8 h period. A dBA is a technical exposure term representing an average effect of sound pressure as perceived by the ear, i.e. not the actual distribution of sound pressures from the sound spectrum. Sound exposure meters are generally calibrated in these units. Like radioactivity, shorter exposure times permit higher dBA levels. The simplest course of action is that those laboratories which employ sonicators should refer to the latest current industrial practice in their own country.

There are three ways in which workers can be protected against radiated noise:

(1) Generally, any person working with ultrasonic equipment should wear either acoustic ear muffs or in-the-ear protectors. These are generally available for use in noisy industrial environments. The simple ultrasonic cleaning bath is not normally sufficiently noisy to warrant more than this simple precaution.
(2) It is good practice to isolate ultrasonic equipment in laboratories where access is limited — a closed door is a reasonable sound barrier.
(3) It is recommended that an acoustic screen is placed around the ultrasonic equipment, particularly in the case of powerful probe systems. This can take the form of a box lined with a sound-absorbing material, say expanded polystyrene or polythene bubble wrapping. In the chemical laboratory a fume cupboard can be partially soundproofed in this way. A neat solution to this problem, particularly for teaching laboratories, is illustrated in Fig. 1.13 where the soundproof box containing a probe system is mounted over storage space for spares. The whole unit is on wheels so that it can be moved around the laboratory.

REFERENCES

[1] B. Brown and J. E. Goodman, *High intensity ultrasonics*, Iliffe Books Ltd, London, 1965.
[2] I. E. El'Piner, *Ultrasound: physical, chemical and biological effects*, Consultants Bureau, New York, 1964.
[3] K. S. Suslick, *Ultrasound: its chemical, physical and biological effects*, VCH, 1988.
[4] T. J. Mason and J. P. Lorimer, *Sonochemistry: theory, applications and uses of ultrasound in chemistry'*, Ellis Horwood, 1989.
[5] Chemistry with Ultrasound, Ed. T. J. Mason, *Critical Reports in Applied Chemistry No 28*, Society for Chemical Industry, 1990.
[6] *Advances in sonochemistry*, Volume 1, Ed. T. J. Mason, JAI Press, 1990.
[7] *Sonochemistry*: *the uses of ultrasound in chemistry*, Ed. T. J. Mason, Royal Society of Chemistry, 1990.
[8] S. V. Ley and C. M. R. Low, *Ultrasound in Synthesis*, Springer-Verlag, 1989.
[9] Special Edition, *Ultrasonics* (1987) **25** (1), January.
[10] *Ultrasonics International 89, Conference Proceedings*, (1990), Butterworths.
[11] Special Edition, *Ultrasonics* (1990) **28** (5), September.
[12] K. S. Suslick, *Scientific American* (1989) 62.
[13] M. A. Margulis, 'The nature of sonochemical reactions and sonoluminescence', in *Advances in sonochemistry*, ed T. J. Mason, Vol 1, JAI Press, UK (1990).

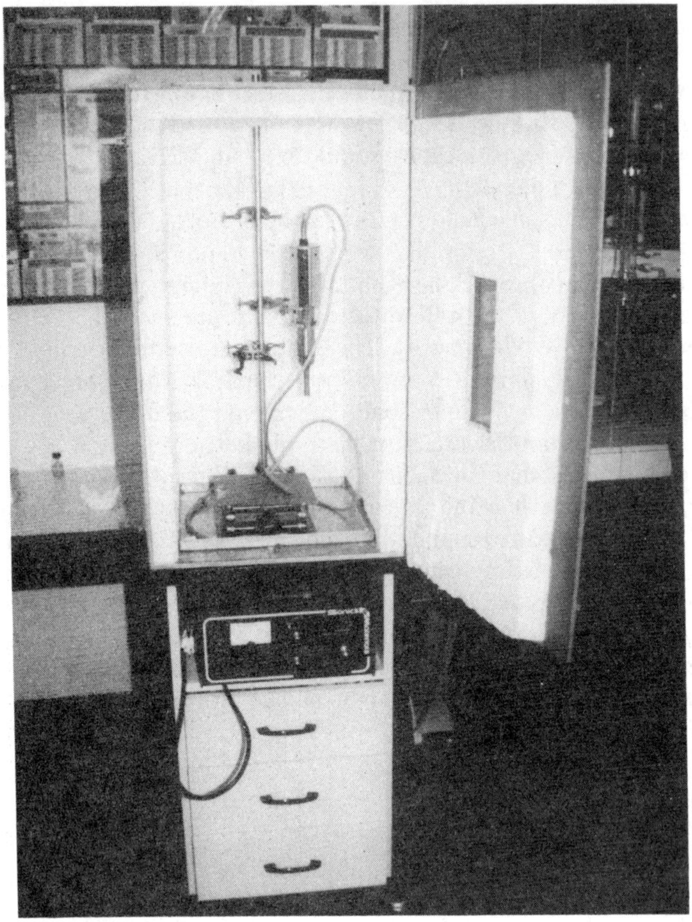

Fig. 1.13 — Movable soundproof cabinet for a probe system as used in the laboratories of J. Berlan in Toulouse.

[14] G. Cum, R. Gallo, A. Spadaro and G. Galli, *J. Chem. Soc. Perkin Trans. 11*, (1988) 376.
[15] J. C. de Sousa-Barboza, C. Petrier and J. L. Luche, *J. Org. Chem.* (1988) **53**, 1212.
[16] P. Riesz, T. Kondo and C. Murali Krishna, *Ultrasonics* (1990) **28**, 295.
[17] P. Kruus, M. O'Neill and D. Robinson, *Ultrasonics* (1990) **28**, 304.
[18] G. Price, 'The use of ultrasound for the controlled degradation of polymer solutions', in *Advances in Sonochemistry*, ed T. J. Mason, Vol 1, JAI Press (1990).
[19] J. D. Sprich and G. S. Lewandos, *Inorg. Chim. Acta* (1982) **76**, 1241.
[20] J. Lindley, P. J. Lorimer and T. J. Mason, *Ultrasonics* (1986) **24**, 292.
[21] R. S. Davidson, A. Safdar, J. D. Spencer and D. W. Lewis, *Ultrasonics* (1987) **25**, 35.

[22] K. S. Suslick and R. E. Johnson *J. Am. Chem. Soc.* (1984) **106**, 292.
[23] R. Walker, *Chemistry in Britain* (1990) **26**, 251.
[24] T. Ando and T. Kimura, Ultrasound in Solid Supported Chemical Reactions, in Advances in Sonochemistry, ed T. J. Mason, Vol 2, JAI Press, (1991).
[25] R. L. Hunicke, *Ultrasonics* (1990) **28**, 291.
[26] R. E. Verral and C. M. Sehgal, Sonoluminescence, in *Ultrasound: its chemical physical and biological effects*, ed K. S. Suslick, VCH, Weinheim, 1988.
[27] A. A. Atchley and L. A. Crum, 'Acoustic cavitation and bubble dynamics', in *Ultrasound: its chemical physical and biological effects*, ed K. S. Suslick, VCH, Weinheim, 1988.
[28] E. G. Janzen, *Acc.Chem.Res.* (1971) **4**, 31.
[29] E. G. Janzen, *Free radicals in biology*, Chapter 4, 'A critical review of spin trapping in biological systems', ed. W. A. Pryor (1980) Academic Press, NY.
[30] G. P. Buettner, *Free Radicals Med. Biol.* (1987) **3**, 259.
[31] P. Riesz, T. Kondo and C. M. Krishna, *Free Rad. Res. Comm.* (1990) **10**, 27.
[32] P. Riesz, T. Kondo and C. M. Krishna, *Ultrasonics* (1990) **28**, 295.
[33] N. Miller, *Trans Faraday Soc.* (1950) **46**, 546.
[34] J. A. LaVerne and R. H. Schuler, *J. Phys. Chem.* (1987) **91**, 5770.
[35] This experimental description was provided by F. Patat, Laboratoire de Biophysique Medicale, Universite Francois Rabelais, 37032 Tours Cedex, France,
[36] R. W. Matthews, *Rad. Res.* (1980) **83**, 27.
[37] M. Anbar, D. Meyerstein and P. Neta, *J.Phys.Chem.* (1966) **70**, 2660.
[38] J. R. McLean and A. J. Mortimer, *Ultrasound Med. Biol.* (1988) **14**, 59.
[39] F. Patat, Etude des besoins et methodes ultrasonores applicables aux expériences de microgravité, Thèse de Doctorat, Universite de Pierre et Marie Curie, Paris 6 (1984) (in French).
[40] The full description of the exposure tank was provided by Professor P. Riesz, Radiation Oncology Branch, National Cancer Institute, 9000 Rockville Pike, Bethesda, Maryland 20892, USA.
[41] A. J.Carmichael, M. M. Mossoba, P. Riesz and C. L. Christman, *IEEE Trans. Ultra. Ferr. Freq. Cont.*, UFFC-33(2), 148, March 1986.
[42] R. D. Corsaro, J. D. Klunder and J. Jarzynski, *J. Acoust. Soc. Am.* (1980) **68**, 655.
[43] J. C. Lockwood and J. G. Willette, *J. Acoust. Soc. Am.* (1973) **53**, 735.
[44] T. G. Muir and E. L. Carstensen, *Ultrasound Med. Biol.* (1980) **6**, 345.
[45] P. Riesz, D. Berdahl and C. L. Christman, *Envir. Health Pers.* (1985) **64**, 233.
[46] The rate of iodine liberation from KI solution can be enhanced by a factor of five in the presence of a small amount of CCl_4, and this method can be used to demonstrate the process (see Chapter 2). The enhancement is the result of sonochemical generation of additional radicals from the haloalkane, which liberates iodine more readily from the KI. The results from this method are thus a reflection of the total number of radicals produced during insonation and are not restricted to those generated from water alone (see A. Weissler, H. W. Cooper and S. Snyder, *J. Acoust. Soc. Am.* (1948) **20**, 589).

2

The ultrasonic cleaning bath

The simple ultrasonic cleaning bath is by far the most widely available and cheapest source of ultrasonic irradiation for the chemical laboratory. It is for this reason that so many sonochemists begin their studies using cleaning baths.

2.1 BATH CONSTRUCTION

The construction of an ultrasonic cleaning bath is very simple — a laboratory model generally consists of a stainless steel tank of rectangular cross-section with transducers firmly attached underneath the flat base (piezoelectric transducers are normally bonded with epoxy). Some tanks, particularly the larger sizes, also have some form of thermostatted heater. A few modern laboratory-scale models have adjustable power but these are in a minority (Fig. 2.1). The frequency and power of an ultrasonic bath depends upon the type and number of transducers used in its construction. Generally the ultrasonic power available in a bath using modern piezoelectric transducers is of quite low intensity (of the order of 1–5 $W\,cm^{-2}$), with an operating frequency of approximately 40 kHz. There are a few variants on this design in that some baths operate at different (but fixed) frequencies. Some cleaning baths also operate with a 'frequency sweep' of a few kiloHertz about a central frequency (e.g. Decon models [1] operate at 40 ∓ 5 kHz). The sweep is designed to avoid standing waves in the tank. Every part of an article for cleaning is thus subjected to the cavitation zone, since the nodes in the waves alternate in height from the base. However, this is not regarded as a major benefit by all manufacturers.

The smallest baths suitable for sonochemistry are of about 1.5 l capacity and a single transducer is sufficient to drive this with a power rating of around 50 W. Larger systems up to 50 000 litres are commercially available and employ an array of transducers on the base to introduce uniform power density into the liquid. One large bath produced by the Lewis Corporation [2] in the USA, for example, has a base area of 15×3 ft (4.5×0.9 m) which employs six clusters of 30 magnetostrictive transducers and has an operational power of 1.5 kW.

A real problem when trying to reproduce work reported in the literature is that ultrasonic baths operate at frequencies and powers dependent upon the transducers

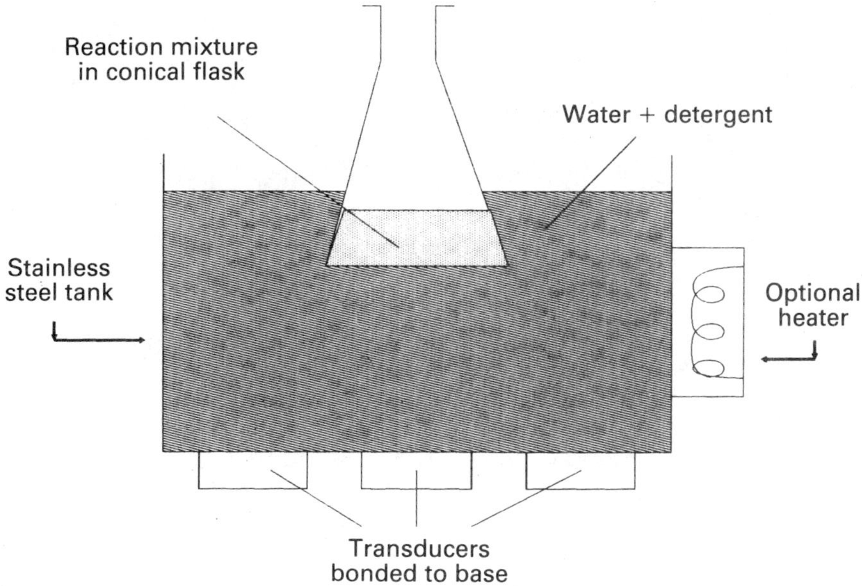

Fig. 2.1 — Schematic diagram of an ultrasonic bath.

employed and with geometries which are specific to the particular manufacturer. This means that it is difficult to make direct comparisons between studies done in different makes (or models) of ultrasonic bath.

Water containing detergent is normally used as the coupling medium in the bath, and this will limit the upper temperatures of operation to below 100°C. A change in medium will allow a greater range of temperature to be studied but this will also result in a change in energy transfer from transducer to the contents of the reaction vessel. Temperature control in baths around ambient temperatures is poor because the passage of power ultrasound through the bath liquid generates a small amount of heat. The system can be operated under thermostatic control at temperatures above that which is reached on thermal equilibration of the bath under its normal working conditions. It is also vitally important to record the temperature *inside* the reaction vessel during sonication since this generally tends to be a few degrees above that of the bath itself owing to ultrasonic heating through the walls of the immersed reaction vessel. Temperature control is dealt with in detail in section 2.3.2 below.

2.2 CHOOSING THE CORRECT TYPE OF ULTRASONIC BATH

Not all commercially available ultrasonic cleaning baths are suitable for sonochemistry. The ideal bath must generate sufficient power to promote cavitation within the reaction vessel when it is immersed in the bath, i.e. enough power to penetrate the walls of the reaction vessel. Naturally the bath must also be large enough to accommodate the reaction vessels used.

If there is a bath available in the laboratory it can easily be established whether it is suitable for sonochemistry by subjecting it to the foil test. The bath should first be filled with water containing some 1 or 2% detergent — in our laboratory we use Decon 90 — the detergent permits much more even cavitation in the bath water. Indeed when the bath contains just tap water a distinct change in 'note' can be detected as the detergent is added (see section 2.5.1). A piece of ordinary aluminium kitchen foil about six inches (15 cm) square should be dipped into the bath (DO NOT IMMERSE FINGERS). After 30 s remove the foil and if it is liberally perforated then the bath has passed its test and is powerful enough for sonochemistry.

2.3 SETTING UP ULTRASONIC CLEANING BATH FOR SONOCHEMISTRY

Although it is possible to use the bath itself as a reaction vessel this is seldom done because of the problems involved in corrosion of the bath walls and containment of any evolved vapours and gases. Normal usage, therefore, involves the immersion of standard glass reaction vessels into the bath [3]. This is of great importance, since conventional glass apparatus can simply be transferred into the bath: this means that an inert atmosphere or a static pressure can be readily achieved and maintained throughout a sonochemical reaction which involves an ultrasonic bath.

2.3.1 Positioning the reaction vessel

Do not assume, however, that all that is required for sonochemistry experiments is to simply immerse your reaction vessel, switch on and wait for results. First consider the way in which your bath has been constructed. The ultrasonic transducer(s) will, most likely, be attached to the underside of the metallic base of the bath and the most intense power will be found directly above a transducer. However, the ultrasonic power in the bath liquid will not be uniform with distance from the base. This is because ultrasound, like any sound, passes through water in the form of a wave and the wave will have positions of maximum amplitude at multiples of a fixed distance above the source. These distances may be calculated using the simple equation $\upsilon=f\lambda$. For example, since the velocity of sound (υ) through water is approximately 1500 m s^{-1}, for a transducer operating at 40 000 cycles per second (f=40 kHz) the wavelength (λ) of the ultrasound in water will be about 4 cm. We may expect the maximum effect to occur at vertical intervals of 2 cm. This series of horizontal regions of high intensity can be located by sonicating a large piece of aluminium foil placed vertically in the bath. A number of distinct horizontal series of perforations appear after some 30 s immersion (Fig. 2.2). Using a series of foil sheets the power regions of the bath can be quite accurately identified because the maximum perforations occur at maximum intensity.

Naturally the reaction vessel should be located at the point where the maximum sonochemical effect can be achieved. Observation of the surface of the reaction solution during vertical adjustment of vessel depth will show the optimum position by the point at which maximum surface disturbance occurs. For a bath with a single transducer on the base this is a simple task since the reaction vessel must be located vertically above the single transducer. For a bath with multiple transducers the foil

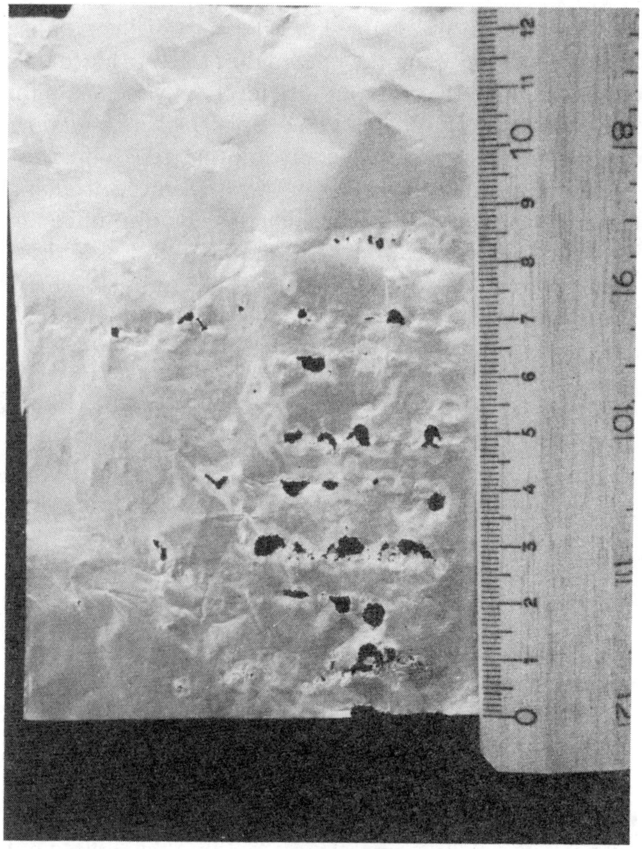

Fig. 2.2 — Perforations in foil on immersion in an ultrasonic bath. Note that the holes appear in distinct horizontal lines at the points of maximum sonochemical effect.

'mapping' technique can be used to locate the regions of most intense power for horizontal positioning.

Another approach to the determination of the way in which the 'sonic energy' is distributed within the bath is the use of thermistor probes [4]. Pugin has shown that the distribution of ultrasonic energy is not at all homogeneous by measuring the local temperature at different places in the sonicated media [5]. Maxima and minima appear, depending on the shape of the vessel, the liquid height and the input power. In some cases standing waves may be formed, but the question as to whether a larger sonochemical effect is obtained with standing waves or in their absence has not been adequately answered.

Once the correct position for the reaction vessel has been identified it is a simple matter to arrange stands and clamps in order to locate it in the required position. Essentially this will act as a template for future reactions since repeated use of the same vessel in the same location in the bath should give reasonably reproducible results.

2.3.2 Temperature control

All ultrasonic baths warm up under the influence of the transducers in the base, and this creates certain problems in terms of both the effectiveness and the reproducibility of the sonochemical results obtained. For reproducible results the bath must be thermostatted either with circulating coolant or by the addition of ice. Both techniques disturb the normal propagation of ultrasound through the bath. The simplest solution to the problem is to determine the 'equilibrium' temperature of the bath (i.e. the maximum temperature which the water attains and maintains under continuous running conditions) and perform most reactions under these conditions. For small baths this is about 45°C and so, before starting any experiments, it is advisable to fill it with hot water so that equilibrium temperature is attained in the shortest time possible.

If it is intended to perform chemical reactions while the bath water is rising in temperature it may be necessary to adjust the position of your reaction vessel occasionally to maintain optimum sonication effect, since the wavelength of ultrasound in water will vary with temperature.

2.4 THE DESIGN OF THE REACTION VESSEL

Once the bath has been chosen, a reaction vessel of the correct design must be chosen. For normal chemical reactions, particularly those involving heat, round-bottomed flasks are utilized, but for sonochemistry in an ultrasonic bath the vessel should be one with a flat-bottom, e.g. a conical flask. The reason for this is that the energy is radiating vertically as sound waves from the base of the bath and this energy has to be transferred through the glass walls of the vessel into the reaction itself. The energy transference is much more effective when the sound impinges directly on the flat base of a conical flask rather than hitting the underside of a spherical container at an angle, since when this happens some energy will be reflected away.

Another important consideration when using baths to perform sonochemical reactions is that it may be necessary to stir the mixture mechanically to achieve the maximum effect of the ultrasonic irradiation. This is particularly important when using solid–liquid mixtures where the solid is neither dispersed nor agitated throughout the reaction by sonication alone and simply sits on the base of the vessel where it is only partially available for reaction. The reasons why additional stirring is so important in such cases is that:

(a) it ensures the reactant powder is exposed as fully as possible to the reaction medium during sonication, and
(b) with stirring there will be no layer of powder on the base of the vessel which could produce considerable attenuation of the incident sound energy.

2.4.1 A practical comparison of vessels for use with ultrasonic baths

A round-bottomed flask was compared with two different-sized conical flasks to illustrate the most appropriate vessel shape for use with an ultrasonic bath [6]. A small single-transducer bath was used in this exercise, a Kerry Pulsatron 60 [7]. This bath operates at 38 kHz, is rated at 50 W and has internal dimensions of 14×15 cm and a depth of 10 cm. In each experiment the flask was positioned for maximum sonochemical effect (see above).

(a) Homogeneous model

As a model for a homogeneous reaction, the liberation of iodine by sonic irradiation of 4% aqueous potassium iodide (50 cm^3, containing 2.5 cm^3 carbon tetrachloride) was chosen. The reaction was monitored by periodically withdrawing aliquots and measuring any increase in the liberation of iodine by increases in its absorption spectrum at 351 nm. The results obtained at 30°C (Fig. 2.3) illustrate the great importance of both the source of ultrasound and the type of reaction vessel chosen for sonochemical studies.

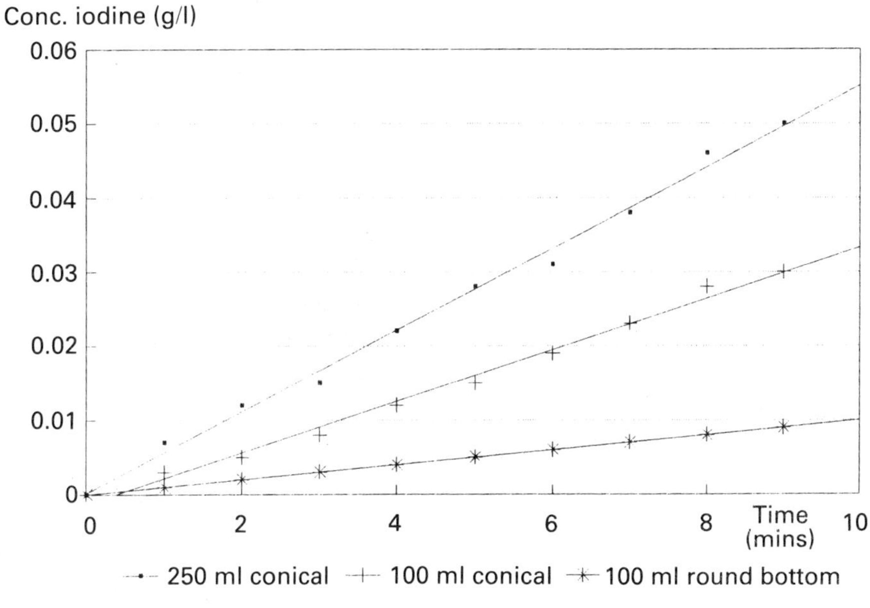

Fig. 2.3 — Dependence of iodine liberation on vessel shape.

For a reaction volume of 50 cm^3 a larger effect is obtained in a 250-cm^3 rather than a 100 cm^3 conical flask. Both vessels present the preferred configuration for transfer of energy from the bath to the reaction media, i.e. a flat glass base at right angles to the propagating sound wave. The results suggest that the greater base area of the 250-cm^3 flask allows more power input to the system, and hence a more effective sonochemical reaction. The use of a 100-cm^3 round-bottomed flask clearly results in a reaction which is inferior to either of these.

(b) Heterogeneous model

The heterogeneous 'reaction' chosen was the reduction in particle size of commercial potassium carbonate (1 g) by sonication in dimethylformamide (50 cm3). Particle sizes were monitored using a Gallai instrument, and the lower limit of particle size reduction was arbitarily taken to be the point at which the mixture became too cloudy for the instrument to function properly (Fig. 2.4).

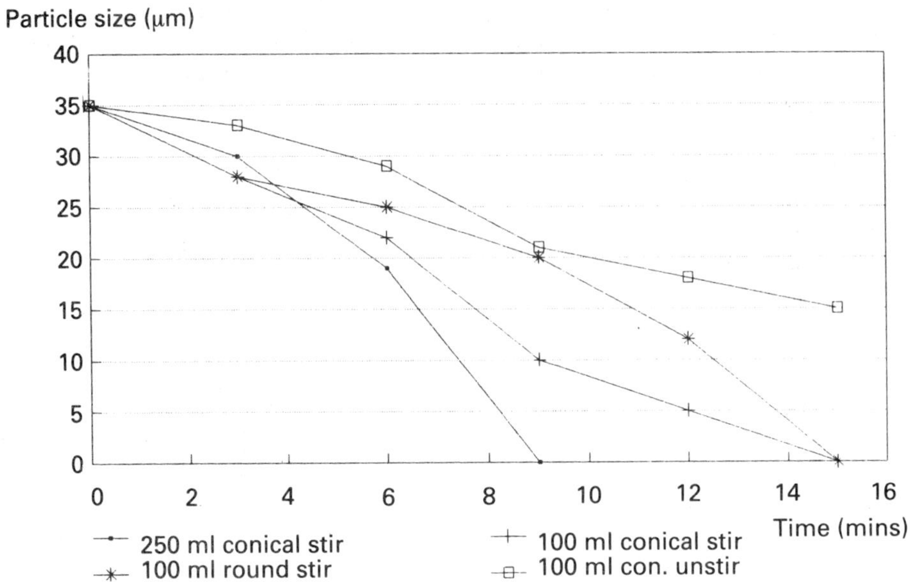

Fig. 2.4 — Dependence of size reduction on vessel shape.

Particle fragmentation using a 100-cm^3 vessel makes it clear that to achieve optimum results mechanical stirring of the heterogeneous mixture is vital. The trend in results for particle size reduction in differently shaped vessels is the same as for the homogeneous reaction, in that the flat-bottomed conical was more effective than the round-bottomed flask.

From these results we can conclude that the flat bottom of a conical flask is the most effective shape for use in an ultrasonic bath but that the best results are obtained with vessels where there is a large base area with respect to volume. Clearly stirring is essential for solid/liquid heterogeneous reactions.

2.5 LABORATORY DEMONSTRATIONS OF THE PHYSICAL EFFECTS OF ULTRASOUND

The short experiments described below have been developed as demonstrations which have been used in lectures and seminars on sonochemistry. They are all of short duration and mainly show aspects of the mechanical effects of power ultrasound. In these examples the bath can be used from cold — no special heating or thermostatting is required.

2.5.1 The effect of added detergent

Fill the bath with pure distilled water and turn it on. Note the audible sound emitted from the tank owing to metal vibrations and the sub-harmonic content of the ultrasonic source. The surface is often subject to erratic disturbance. The addition of about 1–2% detergent changes the note of the sound and causes a more even and enhanced disturbance of the surface.

The detergent lowers the surface tension of the water which 'lubricates' the difficult process of pulling the water molecules apart and thereby facilitates cavitation. This gives a much more efficient transfer of ultrasonic energy through the water — and hence into the reaction vessel. In technical terms the detergent has produced much better coupling between the ultrasonically reverberating base of the tank and liquid medium contained in it. The bath has become a much more reproducible ultrasonic source.

2.5.2 Ultrasonic degassing

The cavitational effects which are the basis of sonochemical action are also the reason for the extremely effective use of ultrasound to degas liquids. Any dissolved gases or gas bubbles in the medium act as nuclei for the formation of cavitation bubbles. Such bubbles are not easily collapsed in the compression cycle of the wave owing to the fact that they contain gas, and they will continue to grow on further rarefaction cycles, filling with more gas and eventually floating to the surface. Since the rarefaction cycles are taking place extremely rapidly (around 40 000 times per second using an ultrasonic bath) the bubbles grow so quickly that degassing appears to occur almost instantaneously.

To produce a gasified liquid place some granular zinc metal in 4-M hydrochloric acid (125 cm^3) in a 500 cm^3 conical flask and allow the effervescence to start; swirl occasionally. After about 30 s the liquid will become opaque with liberated hydrogen and should be diluted with a further 125 cm^3 water to reduce the hydrogen liberation. Now place the base of the flask onto the surface of the water in the ultrasonic bath. As if by magic the cloudiness is cleared. The process can be repeated as long as the zinc remains and the degassing can be seen to occur as a front which rises from the base.

It is also worth noting that the production of the gas at the metal surface also changes from the very large number of tiny bubbles which generate cloudiness to larger bubbles in the ultrasonic field which rapidly rise to the surface leaving the liquid clear.

Ultrasonic degassing has found applications in many areas. In its simplest form it is used to degas solvents for use in high-performance liquid chromatography (HPLC). The electroplating of metals invariably involves gas generation at one or other of the electrodes. The physical presence of the gas acts as a barrier to efficient passage of current (discharge of ions); ultrasound can improve electroplating and indeed general electrochemical processes by removing this gas 'barrier' [8]. A somewhat different development involves use of ultrasonic equipment for the removal of volatile constituents in crude oil fractions, i.e. the 'degassing' of low boiling material such as methane and ethane [9].

2.5.3 Reactions involving metal surfaces

There are two types of reaction involving metals: one in which the metal is a reagent and is consumed in the process and a second in which the metal functions as a catalyst. Metals which are ultrasonically irradiated in a solvent suffer considerable surface damage. This effect is caused by two possible processes: (i) the implosion of cavitation bubbles formed from seed nuclei on the surface or (ii) microstreaming of a jet of solvent onto the surface when a cavitation bubble collapses close to it. The result of this pitting is to expose clean new metal surface to the reagents and to increase the effective surface area available for reaction.

The perforation of foil (see above) is a perfect example of this effect. Simply take a clean piece of aluminium kitchen foil about some 15×5 cm and dip it vertically into the sonicated bath liquid and keep it still. Remove the foil after 30 s and note the perforations which have appeared (an overhead projector is excellent for showing these). If the foil was held reasonably steadily in the bath then the perforations will be clearly seen to be in horizontal rows corresponding to the areas of maximum cavitational effect at half-wavelength distances vertically through the bath liquid.

2.5.4 Reactions involving powders

When cavitational collapse occurs on the surface of a particle of solid powder in a solvent the shock wave generated is of sufficiently high energy to fragment that particle, even if it is a metal. The resulting particle size reduction of the powder increases the surface area of the material and consequently its chemical reactivity. This can be demonstrated using blackboard chalk (Fig. 2.5).

Place a few pieces of the chalk in water (150 cm^3) in a 250 cm^3 conical flask. When the flask is dipped into the ultrasonic bath clouds of chalk are seen to be produced from the surface of the solid. Very quickly the liquid becomes cloudy, illustrating the way in which ultrasound can erode surfaces and break down particles to a much smaller size.

2.5.5 Emulsion reactions

Ultrasound is known to generate extremely fine emulsions from mixtures of immiscible liquids. One of the main consequences of these emulsions is the dramatic increase in the interfacial contact area between the liquids, i.e. an increase in the region over which any reaction between species dissolved in the different liquids can take place.

This can be shown by placing some water (100 cm^3) and toluene (50 cm^3) in a 250 cm^3 conical flask and dipping the base of the flask in the ultrasonic bath. By carefully adjusting the depth of immersion the clearly defined phase boundary between the two immiscible liquids is first seen to become agitated. On closer inspection this disturbance can be seen as a large number of tiny 'explosions' at the interface effectively sending small jets of liquid from one phase into the other. Soon a

(a)

(b)

Fig. 2.5 — Chalk (a) before and (b) after 10-s sonication.

cloudiness appears in the interfacial region, and within 1 min substantial emulsification will have been caused. The demonstration is made even more dramatic if some dye is added to make one of the phases brightly coloured (Fig. 2.6).

2.5.6 The wavelength of sound in water

This experiment uses a suspension of copper bronze (finely divided copper with particle sizes around 50 to 100 μm, 1 g) in water containing 1% detergent (100 cm^3). The mixture is shaken and poured into a 100-cm^3 measuring cylinder to give a uniformly bronze-coloured column of liquid. When the base of the cylinder is placed in the ultrasonic bath liquid the uniform suspension changes to a set of horizontal bright metallic lines at separated distances of half the wavelength of the ultrasound in water (for 35-kHz ultrasound the wavelength of sound in water is about 2 cm) (Fig. 2.7). These striations are produced because the copper bronze particles collect at the nodes of the sound waves in the water. This demonstrates a potentially useful application of ultrasound as a method of manipulating small particles in a liquid medium [10].

2.5.7 Homogeneous generation of iodine

Power ultrasound is capable of generating hydrogen peroxide from the fragmentation of water (see Chapter 1). The peroxide will oxidize any iodide ions in solution to produce iodine. This can be visualized by its reaction with starch to produce the characteristic blue colour. Since the power of an ultrasonic bath is not great, sometimes with aqueous KI there is no colour generation. The effect can be enhanced by adding a few drops of carbon tetrachloride to the KI solution.

Place 5% aqueous KI solution (100 cm^3) in a 250 cm^3 conical flask and add a few drops of CCl_4 followed by soluble starch indicator (2 cm^3). The flask is then immersed in the ultrasonic bath to a depth where the clear solution suffers maximum disturbance. Within 1 min the blue starch/iodine colour will start to appear and this will deepen as the reaction proceeds.

2.5.8 A comparison of thermal excitation with sonochemistry

This demonstration was originally designed for a television programme which required a simple reaction to show how the ultrasonic bath might be used in place of a bunsen burner [11].

Sodium thiosulfate (200 cm^3, 0.001 M) and potassium iodide (100 cm^3 0.2 M) are mixed and cooled to 0°C. A solution of potassium persulfate (200 cm^3, 0.2 M) is also cooled. Fifteen cubic centimetres of the cooled thiosulfate solution is mixed with 0.5 cm^3 starch indicator, and two drops of carbon tetrachloride in a 250-cm^3 conical flask. When 4 cm^3 of the persulfate is added the reaction starts. Persulfate slowly oxidizes the iodide ion to iodine, which initially is reduced by the thiosulfate present. The end of the reaction is signalled by the generation of deep blue colour when all of the thiosulfate is consumed.

With cooled solutions and under normal conditions the reaction time is 1 min. When heated with a bunsen on tripod and gauze the time is reduced to about 30 s and in an ultrasonic bath it is also about 30 s.

(a)

(b)

Fig. 2.6 — Toluene/blue water (a) before and (b) after 10-s sonication.

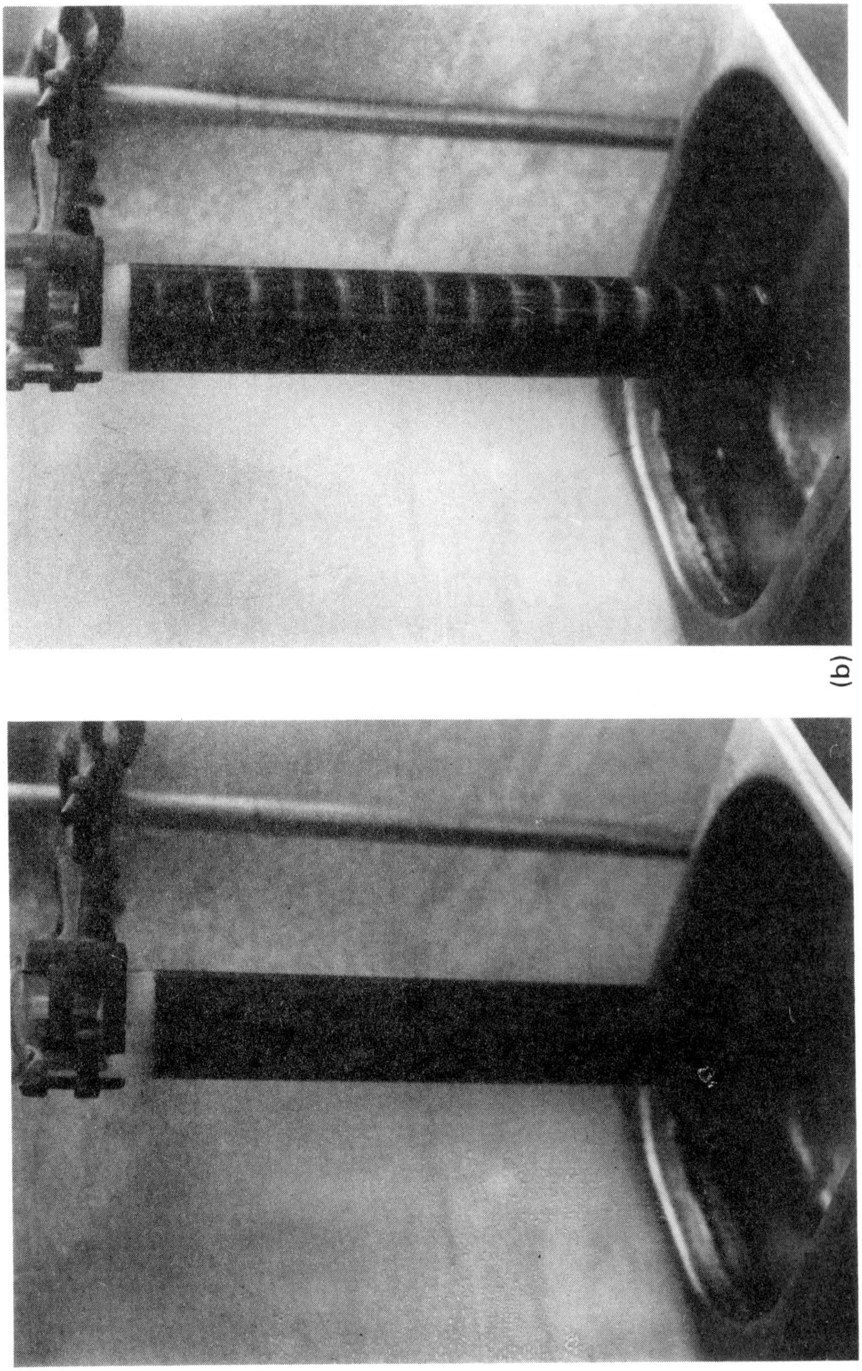

(b)

(a)

Fig. 2.7 — The effect of standing waves on the distribution of a copper bronze suspension. (a) without and (b) during sonication.

2.5.9 Sonochemically enhanced chemiluminescence

This demonstration requires the following solutions: luminol ($0.8\,g\,l^{-1}$), sodium carbonate ($20\,g\,l^{-1}$), hydrogen peroxide (6% or 20 volume), carbon tetrachloride.

In a 100-cm^3 conical flask, luminol solution ($25\,cm^3$) is mixed with an equal volume of aqueous sodium carbonate, carbon tetrachloride ($1\,cm^3$) and 20 drops of peroxide. With all lights turned off the flask should be dipped into the ultrasonic bath. A series of bright sparks of light is emitted from the carbon tetrachloride/water interface as it is agitated by the sonic waves. Light sparks stop on removal of the flask from the tank but are regenerated on dipping in again. This effect can be repeated over a considerable period of time. The chemiluminescent reaction is the result of ultrasonic generation of active (radical) species from vapours contained within the cavitation bubbles when they are subjected to the forcing conditions of cavitational collapse. These free radical species generate light on reaction with luminol (Scheme 2.1).

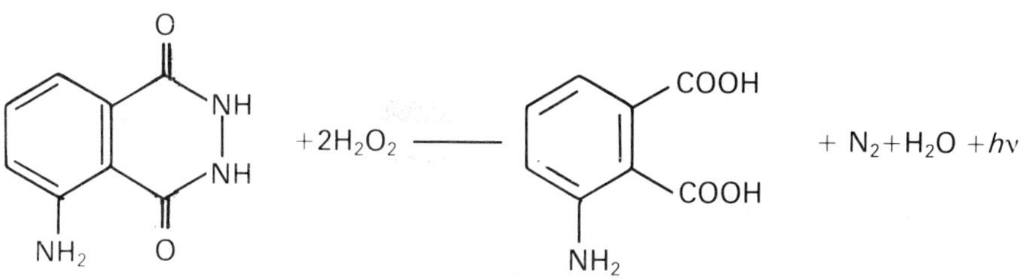

Scheme 2.1.

2.6 ADVANTAGES AND DISAVANTAGES OF USING AN ULTRASONIC BATH FOR SONOCHEMISTRY

Despite the fact that the cleaning bath is the piece of ultrasonic equipment most widely used by chemists it is not necessarily the most effective. The advantages and disadvantages are summarized below.

2.6.1 Advantages

(a) An ultrasonic cleaning bath is certainly the most widely available laboratory source of ultrasonic irradiation.
(b) The sound acoustic field is fairly even throughout the bath liquid.
(c) There is no special adaptation of chemical apparatus required; conventional glassware can be used, which means that the addition of chemicals, an inert atmosphere and the use of reduced and elevated pressures are all easily attainable.

(d) If the cleaning bath itself is used as the reaction vessel larger batch treatment under greater irradiation power can be achieved.

Despite the advantages gained from the use of such a simple piece of apparatus there are a number of considerations which should be borne in mind when using this method of energy input.

2.6.2 Disadvantages

(a) The amount of power dissipated into the reaction from the bath is not very large — usually less than 5 W cm^{-2}.

(b) The energy input should be assessed for each system investigated because power will depend on the size of the bath, the reaction vessel type (and thickness of its walls) and the position of the reaction vessel in the bath.

(c) The ultrasonic frequency at which a cleaning bath operates is not universally the same. Most manufacturers use around 40 kHz, but 20-kHz baths are not uncommon and higher frequencies may be used in some specialist baths. This may well affect results, particularly when attempting to reproduce those reported in the literature.

(d) Temperature control is not easy, since most cleaning baths warm up slowly during operation. This may lead to inconsistent results when working around room temperature or below. Over a prolonged period of use most baths reach an equilibrium temperature. A heater can be used to establish temperatures thermostatted above this equilibrium. Three solutions are available to the problem of operating at room temperature or below: either (i) operate for very short periods during which the temperature can be assumed to remain essentially constant, (ii) circulate cooling water or (iii) add ice. However, if ice is used it must be remembered that solids will alter the characteristics of sonic wave transmission. Whatever method is chosen it must be emphasized that it is the temperature inside the reaction vessel which must be monitored as this is often a few degrees above that of the bath liquid.

(e) A few manufacturers produce cleaning baths with adjustable power, but the majority of cleaning baths do not have any such control. It is possible to place a rheostat between the mains and the bath which will permit simple power control.

(f) Very few cleaning baths have vapour-tight lids. This means that the use of the whole bath as a batch reactor is not possible where an inert atmosphere or a reflux condenser is required.

It is, however, clear from the variety of experiments selected for this text (see Chapter 5) and the large number in the literature that, despite its drawbacks, the simple ultrasonic cleaning tank is extremely useful for sonochemistry.

2.7 OTHER TYPES OF EQUIPMENT RELATED TO CLEANING BATHS

A laboratory cleaning bath may well prove to be capable of providing sufficient energy to promote a sonochemical reaction. Following this discovery it may be that the process requires more rigorous investigation, refining or adapting to a larger scale. Several pieces of equipment are available to extend the capabilities of the bath.

2.7.1 The submersible transducer assembly

For many years the manufacturers of ultrasonic cleaners have offered the submersible transducer as a method of converting any vessel into an ultrasonic cleaning bath (Fig. 2.8). The submersible can be visualized as a number of transducers which have

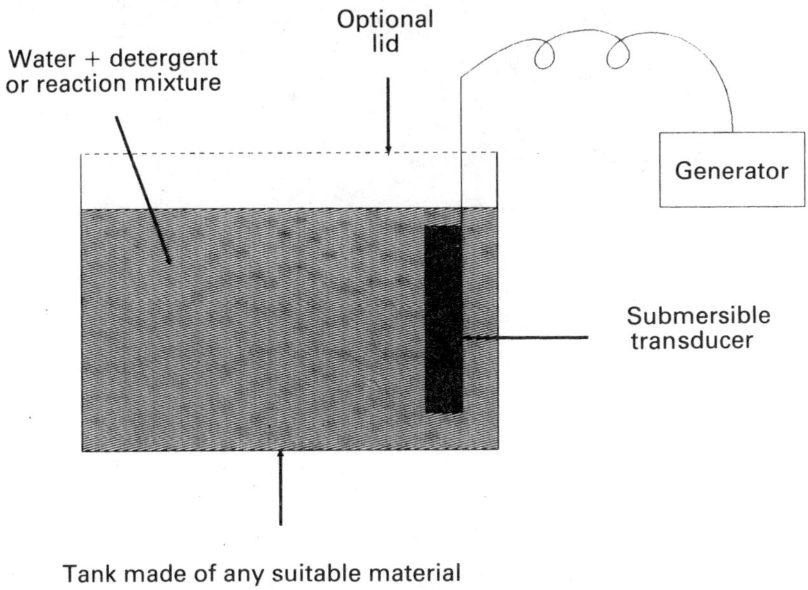

Fig. 2.8 — Schematic diagram of a submersible transducer system.

been sealed into a steel box (Fig. 2.9). These are bonded to a face of the box which will become the radiating face when immersed in a liquid. This system offers several advantages over the cleaning bath.

(a) A container of any size and shape can be converted into an ultrasonic bath. The submersible can be constructed in a variety of sizes.
(b) The submersible transducer can be dipped directly into a chemical reaction, thereby introducing power directly into a system rather than through a vessel wall. Of course this advantage will only be of use if the chemical system does not chemically react with the transducer housing.
(c) The position of the submersible can be adjusted to give maximum sonochemical effect in the particular vessel used.
(d) The vessel to be used can be quite large since any number of submersibles can be used.

The most likely sonochemical applications for submersible transducers will be in the treatment of aqueous or non-corrosive liquid systems.

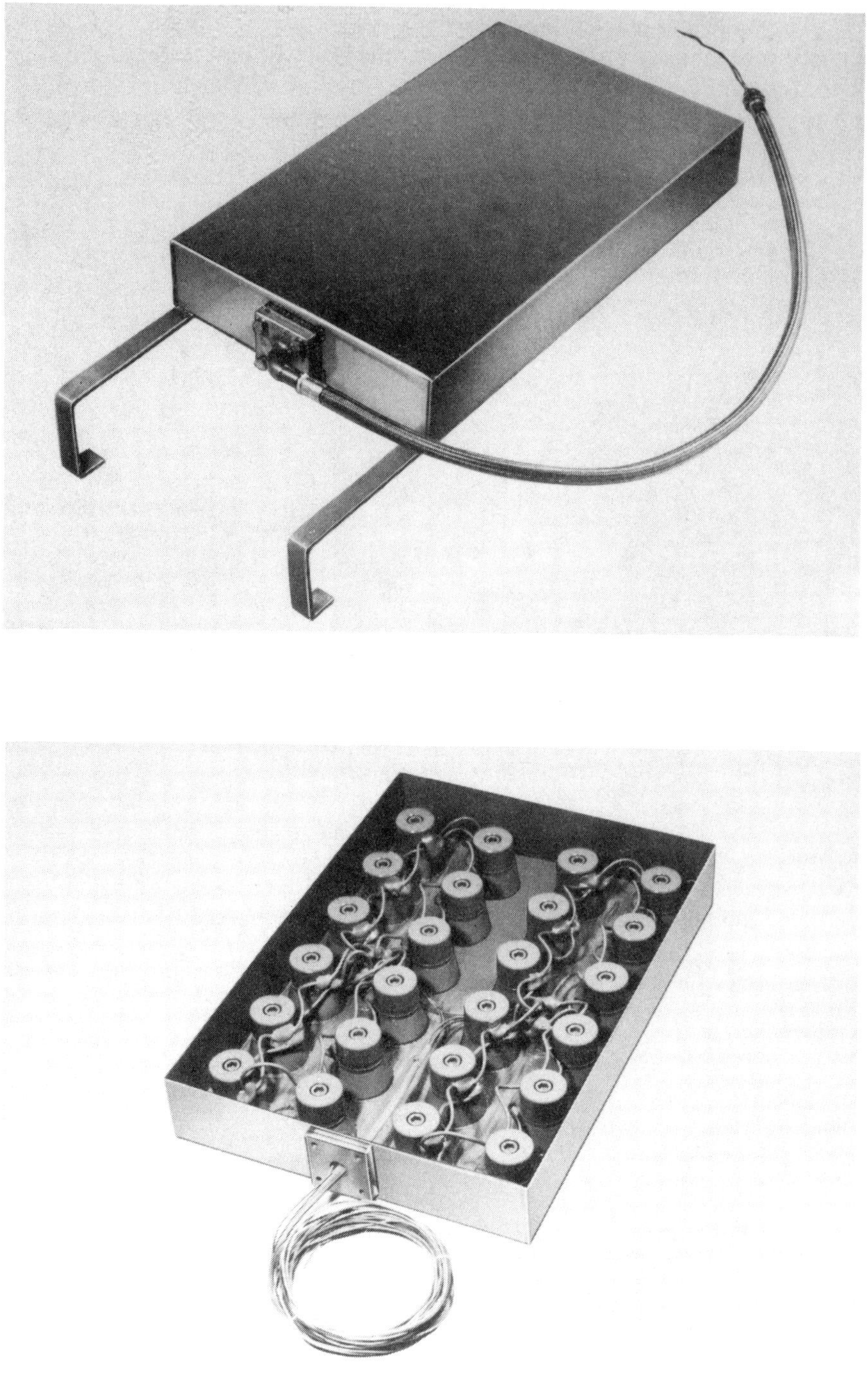

Fig. 2.9 — Construction of a submersible sonication unit showing the array of transducers bonded to one face. (Picture courtesy of Martin Walter Ultraschalltechnik GmbH.)

2.7.2 The cup-horn

If the general utility of an ultrasonic bath could be combined with the controllable power of a probe a useful compromise would be available. Such a compromise system is the cup-horn device (Fig. 2.10), which can be regarded as a small bath into which can be placed the vessel containing the reaction to be sonicated. Originally

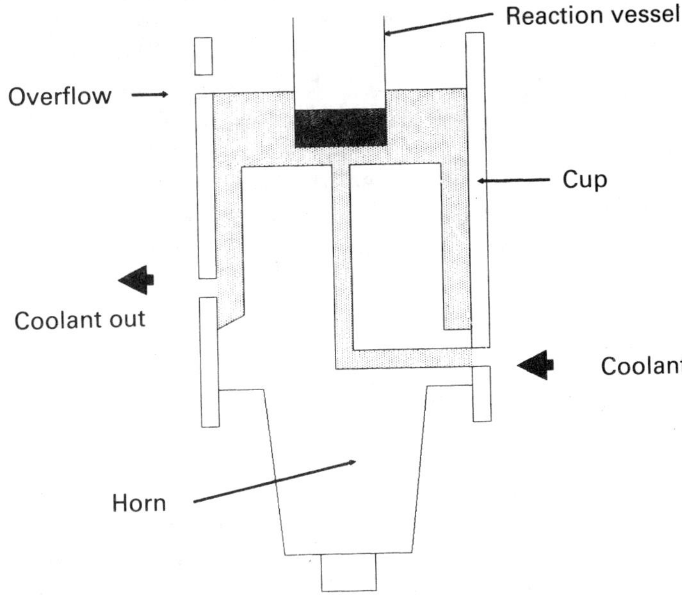

Fig. 2.10 — Schematic diagram of cup-horn.

designed for cell disruption, the use of such a system permits a more quantitative and reproducible study of sonochemical effects than either the cleaning bath or the probe system. The frequency is fixed but the power is tunable, and, unlike the direct immersion sonic horn, no fragments of metal are generated in the reaction itself.

This piece of equipment may be envisaged as a small but powerful ultrasonic bath. In effect it uses a specially designed inverted horn to irradiate a small cup of liquid (normally water) into which the reaction vessel is dipped. The specially designed horn, with a hole drilled through its centre, allows the liquid in the cup to be part of a circulating coolant system which assists in maintaining temperature stability. Since the irradiation source is a modified sonic horn, this system has one major advantage over the cleaning bath — its power can be controlled. The main difficulty with the cup-horn is that it is quite small and this limits the size of the reaction vessel which can be used.

2.7.3 A high freqency ultrasonic bath specially designed for sonochemistry

The ultrasonic bath described below is the type used by A. Henglein of the Hahn-Meitner-Institut, Berlin. Henglein began his studies of the chemical effects of ultrasound as a student studying for a masters degree over 50 years ago and sonochemistry remains one of his research interests [12]. Today most chemists use commercial ultrasonic equipment which operate in the frequency range 20 to 40 kHz, but the studies in Professor Henglein's laboratory are often conducted at a much higher frequency (1 MHz) with the specially developed apparatus described below.

The ultrasound is generated by a quartz or piezoceramic transducer with an emitting face of about 25 cm^2. Fig. 2.11 shows a typical reaction vessel used for sonication of a liquid in an enclosed gas atmosphere. The sound travels through a metal flange of one-wavelength thickness into a bath of cooling water and then moves through the thin glass bottom of the vessel which contains about 30 cm^3 of solution. The thin glass wall of the vessel allows rapid heat exchange between the irradiated liquid and the water bath. The sound is reflected at the liquid surface, where it produces considerable agitation (i.e. a fountain) at higher intensities above 1.5 $W\,cm^{-2}$. The vessel can be moved reproducibly along the vertical axis to locate the point of maximum energy transference as a field of standing waves is established.

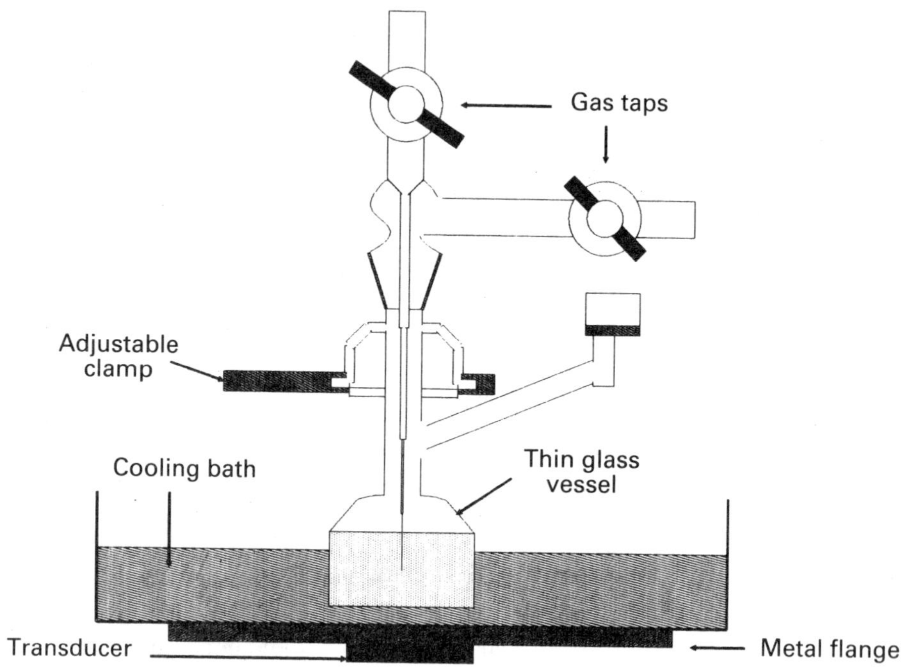

Fig. 2.11 — Cell specially designed for sonochemistry.

An alternative vessel, which is capable of higher insonation intensities, is shown in Fig. 2.12. The glass bottom is a half-wavelength thick and the vessel is placed directly on the metal flange which carries the transducer (a thin film of water between the vessel and the flange ensures efficient acoustic coupling). A significantly higher

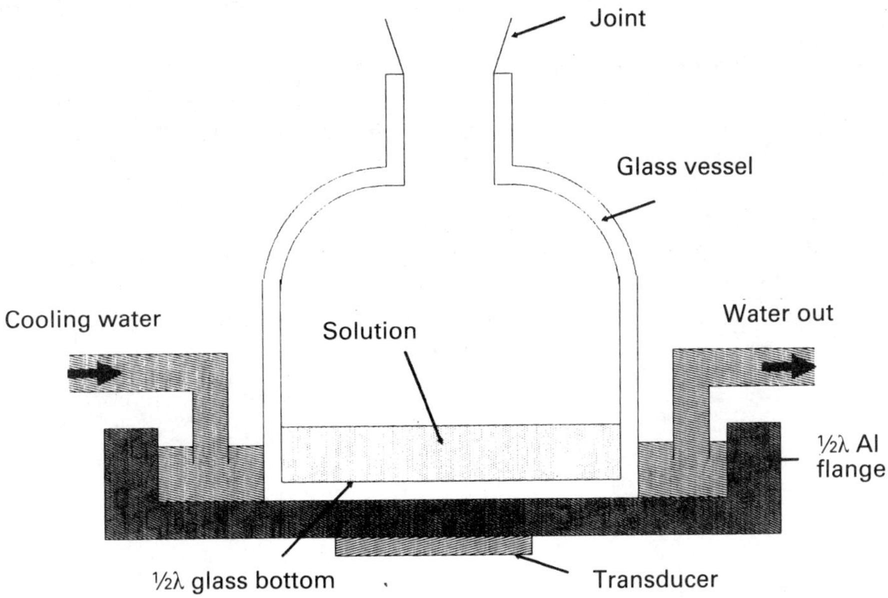

Fig. 2.12 — Powerful high frequency ultrasonic bath appartus.

proportion of the sound intensity emitted by the transducer reaches the solution in this apparatus than in that described above.

2.8 THE POWER OF EQUIPMENT RELATED TO CLEANING BATHS

In any apparatus related to cleaning bath systems the cavitation bubbles are formed as separate entities throughout the insonated liquid medium. This contrasts with the streaming and coalescence of bubbles which occur in the more powerful sonicators, e.g. probe systems. It is therefore difficult to compare the results of experiments carried out with bath and probe types of sonication equipment. There is no doubt, however, that cleaning baths provide sufficient power to identify reactions which are susceptible to sonochemical enhancement and that they are certainly the most convenient to use under laboratory conditions.

REFERENCES

[1] Decon Laboratories Ltd, Conway Street, Hove, Sussex BN3 3LY, UK.
[2] Lewis Corporation, 102 Willenbrock Road, Oxford, Connecticut 06483, USA.
[3] T. J. Mason, J. P. Lorimer and J. Moorehouse, *Education in Chemistry* (1989) **26**, 13.
[4] C. J. Martin and A. N. R. Law, *Ultrasonics* (1983) **21**, 85.
[5] B. Pugin, *Ultrasonics* (1987) **25**, 49.

[6] T. J. Mason, J. P. Lorimer, F. Cuesta and L. Paniwnyk, *Ultrasonics International 89, Conference Proceedings* (1989) 1253.
[7] Kerry Ultrasonics, Hunting Gate Wilbury Way, Hitchin, Herts SG4 0TQ, UK.
[8] T. J. Mason, J. P. Lorimer and D. J. Walton, *Ultrasonics* (1990) **28**, 333.
[9] Ultrasound in Chemical Processing, N. Senapati, *Advances in Sonochemistry* (1991) Vol 2, ed T. J. Mason, JAI Press.
[10] The Manipulation of Particles in an Ultrasonic Field, C. J. Schram, *Advances in Sonochemistry* (1991) Vol 2, ed T. J. Mason, JAI Press.
[11] BBC Television 'The Ultrasonic Bunsen Burner', *Tomorrow's World*, February 14th, 1989.
[12] Sonochemistry — a historical perspective, A. Henglein, *Advances in Sonochemistry* (1992) Vol 3, ed T. J. Mason, JAI Press.

3

The ultrasonic probe

The major disadvantages of the cleaning bath as a source of ultrasound for sonochemistry are its general lack of power and an inability to control the energy input to the chemical reaction, resulting in inconsistent results. One solution to these problems has been found in a piece of equipment originally developed for biochemists as a biological cell disruptor — the ultrasonic probe. Using this apparatus it is possible to introduce ultrasonic energy in the laboratory which can be up to 100 times greater than can be achieved using a cleaning bath (in terms of watts per square centimetre); on a larger scale the energies can be much greater than this. Modern probe systems are based on piezoelectric transducers and they are generally constructed in a similar manner, as described below.

3.1 ULTRASONIC PROBE CONSTRUCTION

In order to increase the amount of ultrasonic power available to a reaction it is desirable to introduce the energy directly into the system rather than rely on its transfer through the water of a tank and then the reaction vessel walls. The simplest method of achieving this would be to have the face of an ultrasonically vibrating transducer immersed in the reaction solution. Unfortunately the maximum amount of power (vibrational amplitude) generated by the transducer assembly is not very large, and if too much electrical power is fed to the piezoceramic it will fail (see section 1.7.3). It is possible to amplify this vibrational amplitude by attaching a specially designed length of metal rod to the end of the transducer. This rod extension is more correctly termed a sonic horn or velocity transformer, and it not only magnifies the acoustic energy available but also allows the transducer to be kept clear of the chemical reaction since only the tip of the rod needs to be immersed in the liquid. It is the complete assembly of transducer plus horn which is referred to as an ultrasonic probe system. The overall construction of a simple probe system is shown in Fig. 3.1 and Fig. 3.2 is a photograph of a typical laboratory model.

The essential parts of the complete probe system are given below. The terminology has been simplified in order to emphasize the purpose of each component.

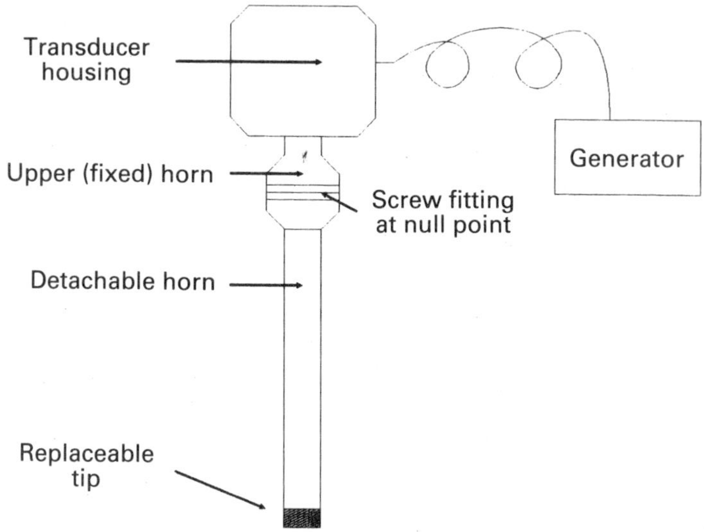

Fig. 3.1 — Schematic diagram of probe system.

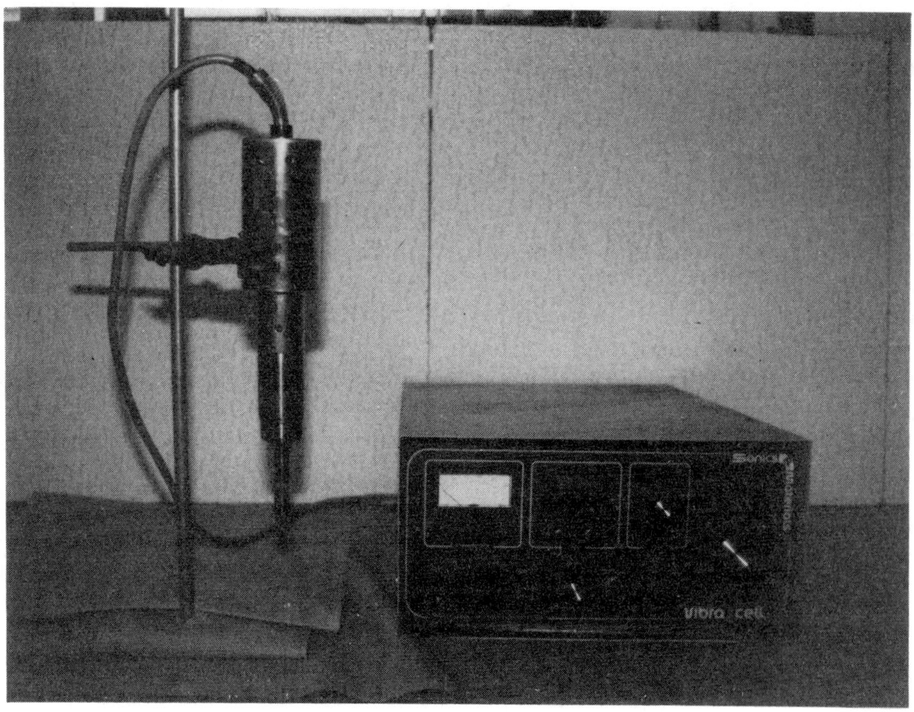

Fig. 3.2 — Typical laboratory probe system.

3.1.1 The generator

This is the source of alternating electrical frequency which supplies the transducer; normally this frequency is 20 kHz. The basic controls on the generator are:

Tuning: For optimum performance a probe system should be 'tuned' before it is used. Tuning is a process whereby the transducer (normally operating at 20 kHz) is brought into resonance as part of the complete probe assembly; this is the point where the minimum power is drawn during operation. This may well be at a frequency a little different from 20 kHz. Thus if a horn is shortened by tip erosion then the resonance frequency will change. Eventually the limits of adjustment of the generator and transducer will be reached and tuning will no longer be possible; when this occurs the only solution is to replace the horn.

Instructions which accompany most commercial probe systems made in the USA (e.g. Branson, Heat Systems, Sonics and Materials [1] suggest that the tuning of the system is best carried out in air at lower power. Technically the system is then said to be not loaded (loading refers to any external change that requires extra power to drive the probe, e.g. when it is immersed in a liquid medium). Tuning in this state ensures that the complete probe system is in resonance. These systems only have a small tolerance to changes in resonance frequency, and sometimes the tuned system will not perform satisfactorily when immersed in the reaction mixture. This is particularly noticeable when operating at low powers. For this reason our probe systems are tuned in the reaction mixture, i.e. under load. If the reaction changes in character during sonication — by warming or altering viscosity, or precipitating a solid, then the power required to maintain optimum resonance changes so that re-tuning may be necessary.

A different type of tuning is often used in probe systems of European manufacture (e.g. Undatim Ultrasonics [2]). These can tolerate a wider range of frequencies for tuning, and the power is optimized for the whole sonication system under load. These systems should, therefore, be tuned when immersed in the reaction. For the Undatim system electrical feedback is incorporated in the design to give continuous adjustment of the tuning as the reaction proceeds.

Power control simply allows the energy to the transducer to be varied — normally on an arbitary 0 to 100% scale — it prevents overload and fracture of the piezoceramic element and reduces metal fatique in the horn.

A *Pulse* facility consists of a timer which switches the power to the probe on and off repeatedly. The 'off' time allows the chemical system to cool between the pulses of sonication. The 'on' time is represented as a fraction of the total time involved in the cycle (about 1 s) i.e. 100% is continuous sonication while 25% represents 0.25-s bursts of power interspersed with 0.75-s 'off' time. This type of pulse operation should not be confused with 'duty cycle' as used in medical ultrasound. In medical parlance the duty cycle refers to the on/off ratio for scanning which involves the emission of pulse of extremely short duration (e.g. 10^{-5} s), containing only a few cycles of ultrasound in the megahertz range. It is in the longer off period that the echos are detected. As a comparison, a chemical sonicator (20 kHz) pulse cycle of 50% will involve many thousand of cycles (less than 10 000 because the start of the cycle is not instantaneous).

3.1.2 The transducer element

The sandwich transducer element is protected by a casing (the transducer housing) which is perforated to allow cooling and may sometimes contain a small fan to help guard against overheating.

3.1.3 The upper (fixed) horn element

This is a robust piece of titanium alloy which is accurately machined to provide an *end face* (commonly 0.5 inch (1.25 cm) diameter) vibrating at maximum amplitude to which other horns may be attached. Together with the detachable horn this is referred to as the emitter or booster. It is important to recognize that this section is subject to ultrasonic motion which is like that of an elastic band, with the maximum vibration at the lower end balanced by a maximum at the transducer face. Naturally in the centre there will be a point of zero motion — a *null point* — it is here that a screw thread may be located for the attachment of optional ancillary equipment, e.g. a flow cell or a polytetrafluoroethylene (PTFE) sleeve (see below).

3.1.4 The detachable horn

An important part of the assembly is the detachable horn, which allows the vibration of the fixed horn to be transmitted through a further length of metal which can be used to magnify the power delivered to a system. Most modern probe systems are designed to operate with a range of detachable horns, each offering a different magnification of power. Horn design is in fact a very important aspect of ultrasonic engineering. The material used for acoustic horns should have high dynamic fatigue strength, low acoustic loss, resistance to cavitation erosion and also be chemically inert. The most suitable material by far for this purpose is titanium alloy. The cavitation which is the source of physicochemical activation is also the source of a common problem with probe systems — tip erosion — which occurs despite the fact that most horns are made of titanium alloy. One major side-effect of erosion is that there will be a physical shortening of the horn below a half wavelength, which will cause a loss of efficiency. This problem is alleviated by the use of screw-on *tips* for the 0.5-inch (1.25 cm) diameter horn in the form of titanium studs; this eliminates the need for a costly replacement of the whole section. Fig. 3.3 shows the effect of cavitational erosion on a screw-on horn tip.

Let us consider, in turn, the engineering factors which are important in the design of different types of detachable horn.

(a) Length

The wavelength of ultrasound in a material is determined by both the type of material and the frequency of the sound wave. In the case of the type of titanium alloy used for horns, the wavelength for 20-kHz sound is about 26 cm, and this defines absolutely the longitudinal dimension of titanium horns. The minimum length of such a titanium alloy horn is a half wavelength, i.e. 13 cm, which will give an exact mirror of the vibrational amplitude supplied at one end to the other. If the distance between the upper fixed probe assembly and the sample being processed needs to be increased

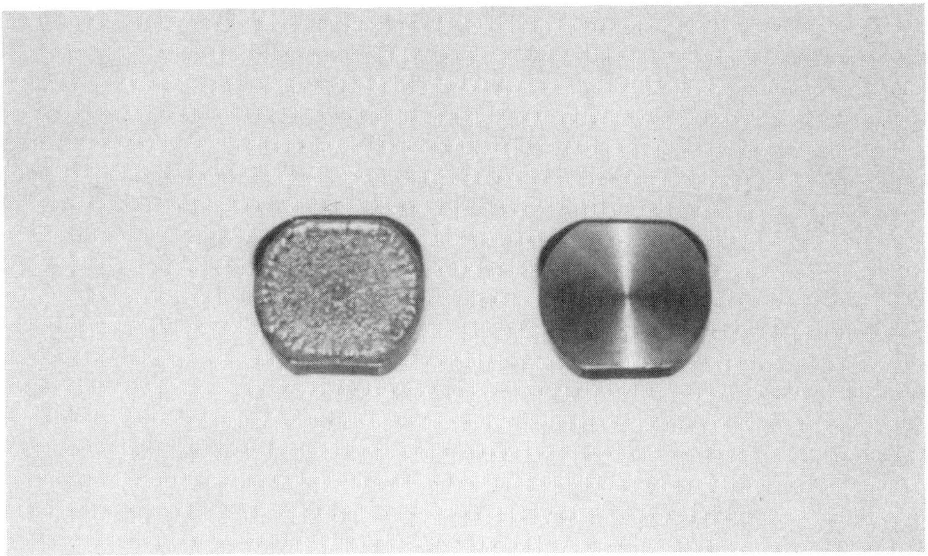

Fig. 3.3 — Cavitational erosion at the detachable tip of a probe system.

this can be done using a horn cut to a multiple of half- wavelengths. This can also be achieved by screwing one half-wavelength horn into another, thereby building up the overall length.

(b) Shape

Four of the most common shapes for horns used with commercial probe systems are shown in Fig. 3.4.

Uniform cylinder: A 13-cm cylinder of titanium alloy will act as an extender for the ultrasonic energy generated by the transducer. It can be used to simply lengthen the working end of the probe system. It will not change the vibrational amplitude.

Linear taper or *cone* If the end diameter of a horn is smaller than the area attached to the driving transducer element then there will be a magnification of the amplitude of ultrasonic vibration (i.e. power). Consider ultrasound entering at one end of the horn at, say, 10 μm peak to peak over an area of 0.5-inch (1.25-cm) diameter. All of this energy must be dissipated at the other end of the horn. If that end is of smaller area, then the vibrational amplitude must be greater. The amplification factor for a linear taper horn is the ratio of the end diameters. The linear taper is the easiest design to manufacture but its potential magnification is normally restricted to a factor of approximately four fold. This is because there will be a great deal of strain developed in the metal of the horn as parts of it undergo large extensions and contractions, particularly towards the working (narrow) end. For both this shape and the exponential horn described below, the null point occurs at approximately 0.4 of the length from the thicker end. This figure is not exact — the precise calculation depends upon the relative diameters of the ends and involves some rather complicated mathematics.

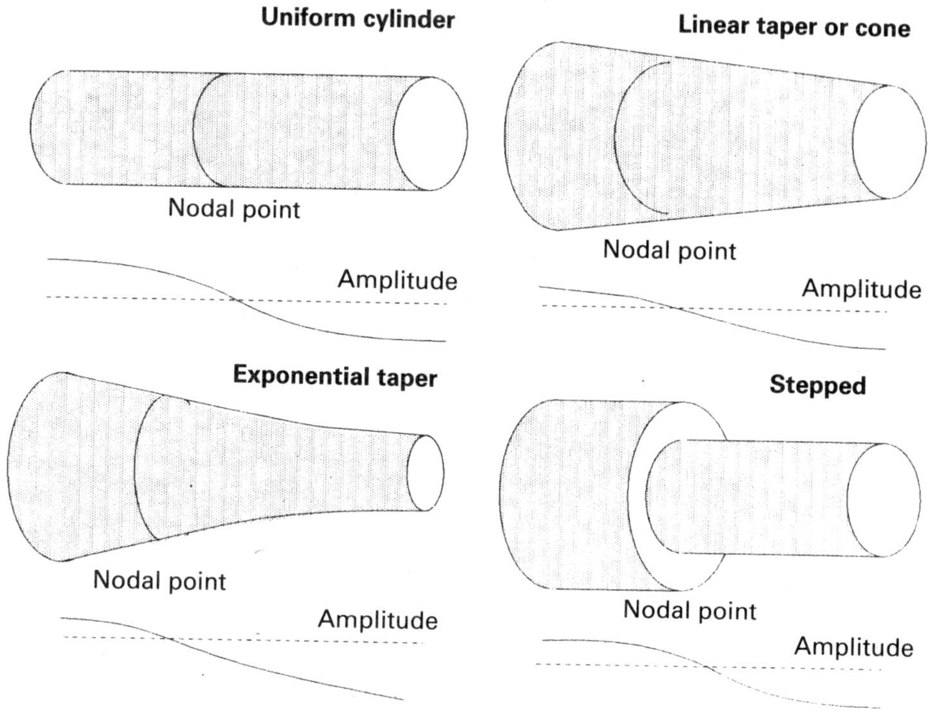

Fig. 3.4 — Horn shapes.

Exponential taper: The exponential taper offers higher magnification factors than the linear taper. Its shape makes it more difficult to manufacture, but the small diameter of the working end (3 mm or less) and its length make it particularly suited to micro applications.

Stepped: There will be no vibration at the midpoint of a 13-cm-long horn fashioned as a uniform cylinder, i.e. this will be a nodal or null point. If this cylinder was to be reduced in diameter at its midpoint to 0.5 of the original cross-sectional area then when vibrational energy was applied to the larger end the smaller end would automatically be subject to a doubling of energy density (energy applied at larger end now emerging through half the area at the smaller end). The 'step' is never as abrupt as shown in the diagram since this would make the horn too fragile. Commercial stepped horns have slightly smoothed shoulders to relieve internal stress.

For a simple stepped design the amplification factor will always be the ratio of the cross-sectional areas, and such horns will accommodate gains of up to 16-fold.

Probe systems are undoubtedly the most efficient method of transmitting ultrasonic energy into a reaction. Although so much extra power is available using the probe system, it is more expensive than the bath and it is significantly less convenient in use, especially in terms of the glassware required. Special seals will be needed if the horn is to be used in any reactions which involve reflux, inert atmospheres or the use of pressures above (or below) ambient.

3.2 CHOOSING A PROBE SYSTEM FOR SONOCHEMISTRY

For the probe system, whatever design of horn is used, a large maximum power density can be achieved at the radiating tip. This can be of the order several hundred watts per square metres, although in many cases much lower energies are required for sonochemistry.

3.2.1 Conventional probe systems

Until quite recently, with the advent of specialist European models (see above), the choice of commercial probe sonicators was somewhat limited. The only frequency available was 20 kHz and the 'extras' were also similar. Essentially the only choice was in colour and shape of the high frequency generator. At the time of writing this is still true of manufacturers of probe systems in the USA: e.g. Branson, Heat Systems and Sonics and Materials. These or any other model should provide the following features for use as laboratory-scale sonicators.

(a) Variable power normally achieved via a dial setting from zero to a maximum output of about 150 W cm^{-2} through a 1.25-cm-diameter probe tip. This will normally approximate to a quoted rating for the instrument of 500 to 600 W.
(b) A pulse facility to allow intermittent irradiation.
(c) A range of interchangeable horns of the types described in section 3.1 above. Replaceable tips (studs) for the 1.25-cm probe.
(d) Horn extenders for the 1.25-cm system with screw-thread fittings at *both* ends. Some extenders have only one end with a screw thread fitting and thus (i) can only be used to extend by a single 13 cm and (ii) cannot be fitted with a replaceable tip.
(e) Flow cells in stainless steel are available, and these allow the probe system to be adapted for continuous processing (see Chapter 4).
(f) Cup-horn attachment for more precise and reproducible (though low power) irradiation.

These probe systems are the 'work-horses' of sonochemistry, requiring only to be tuned to optimum performance in the system under study (see section 3.1.1) before being used to sonicate the system for time periods and at powers which are at the operator's discretion.

3.2.2 Additional desirable features

It has always been possible to have equipment built to one's own specifications and, after commercial equipment has been in use for some time, it is not uncommon to find that other facilities are required. Some of the extra options which are recommended in addition to those above are:

(a) The option to use different frequencies. This will mean using a different transducer and horn for each frequency used and a different frequency generator. The frequency generators can be housed within one box and the sonication treatment can be pre-programmed. There are two commercial systems available which offer a choice of frequencies — from Sonic Systems in the UK and Undatum Ultrasonics in Belgium [1].
(b) Automatic frequency tuning is useful, since the precise frequency match for the probe and the reacting system is not the nominal value for the transducer employed. A read-out of the optimized frequency is then essential.
(c) Some form of feedback to the generator from the transducer element can be used to maintain irradiation power throughout changes which occur during a reaction. A wattmeter reading of power supply to the transducer or vibrational amplitude control of the horn is then an advantage.

3.3 LABORATORY SCALE SONOCHEMICAL REACTORS INVOLVING PROBE SYSTEMS

The simplest method of introducing power ultrasound to a reacting system via a probe would be to dip the probe into the chemical reaction contained in a standard roundbottomed flask or even a beaker. The temptation then would be to turn the power to a maximum and wait to see what wonders of sonochemistry might develop.

3.3.1 The problems associated with using a probe system

Unfortunately the obvious approach of simply dipping the probe tip into a chemical reaction is seldom ideal. Several precautions should be taken when designing a reactor involving a probe as a result of the following factors.

(a) Volatile components may well be forced out of the reaction by the 'degassing' effect of power ultrasound.
(b) Probes can generate a great deal of heat and so cooling may be required.
(c) When high energy ultrasound is introduced into a liquid the power input may produce sufficient power to dampen the effective sound input to a reaction. Thus the *maximum* setting of the generator power control does not necessarily produce the *greatest* sonochemical effect.
(d) Mechanical mixing may be required at low power, although at higher powers the vibrating horn tip does produce sufficient streaming of the liquid away from the tip to produce effective bulk mixing.
(e) Chemical reactions often require vapour-tight apparatus, the slow addition of reagents or inert atmospheres. Any reaction vessel incorporating a standard probe system must therefore be adapted to facilitate such requirements.

3.3.2 Fitting a probe system into standard glassware

Commercial probe systems are supplied with 13 cm horns (the length is dictated by the wavelength of 20-kHz sound in the titanium alloy used). Unfortunately, for most laboratory-scale reactions, this is far too short to enable the vibrating tip to reach the liquid medium. To solve this problem it is necessary to fit a length (or lengths) of titanium alloy rod to the probe and thereby extend the titanium horn to the length required. The next problem is to find a way of sealing the horn into a reaction vessel. Here the difficulty is that at most points in its length the horn is undergoing ultrasonic vibrations. Fortunately, as we have seen above, there is a null point at a quarter-wavelength from the transducer face and multiples of half-wavelengths beyond this first null point. On most commercial sonicators there will be a screw-thread at the first null point (see Fig. 3.1). Anything screwed onto the horn at this point will not be subject to vibration and so it is here that a PTFE sleeve adaptor can be attached and this will push-fit into ground-glass apparatus (Fig. 3.5).

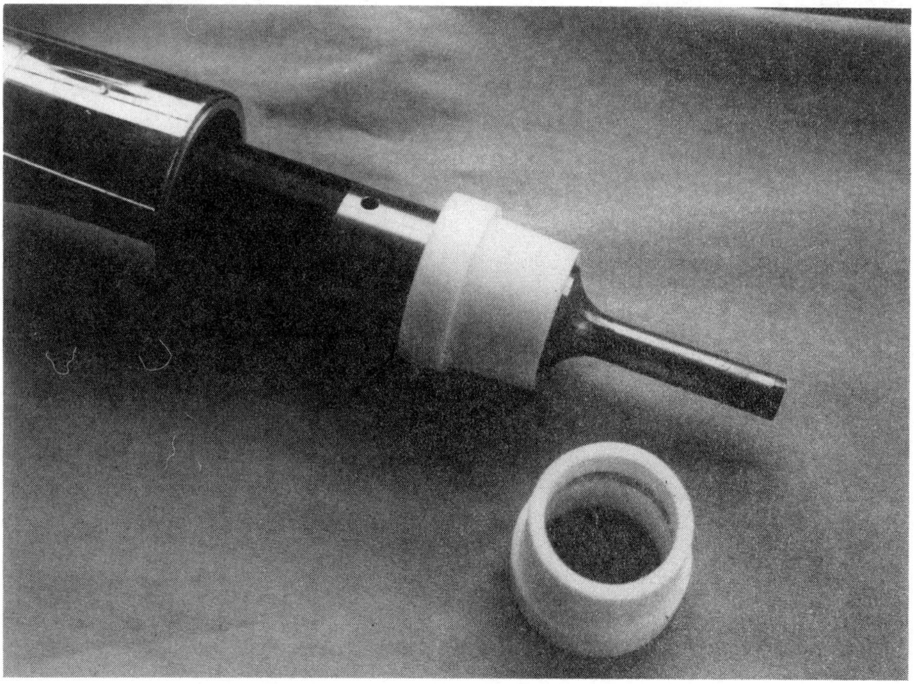

Fig. 3.5 — PTFE sleeve to seal a horn into a standard taper B45 joint.

Using conventional (20-kHz) sonicators the diameter of the screw thread is such that a sleeve for a B34 cone can be made with little difficulty. Unfortunately, because of the diameters involved it is physically impossible to organize a convenient direct seal into a B19 cone.

3.3.3 The design of reaction vessels

A number of solutions to the problems outlined in section 3.3.1 have been proposed. Once the horn has been sealed into the reaction vessel some of the options then available are as follows:

(a) The use of glass beads

A probe system will cause mixing in a reaction as a result of the liquid motion produced by the sonic waves generated at the horn tip. The mixing can be enhanced by putting glass beads in the bottom of the reaction vessel. The beads will be 'bounced' into motion by the sonic waves and thereby cause bulk mixing, but this will also result in a loss of effective energy transfer.

(b) Dimple cell

To facilitate sonic mixing a reaction vessel can be permanently adapted to give more efficient, and reproducible, sonic mixing by providing an indentation on its base (Fig. 3.6). The probe should be positioned vertically above and not less than 1 cm from the indent, which serves to disperse the sonic waves as they impinge upon and are reflected from the base.

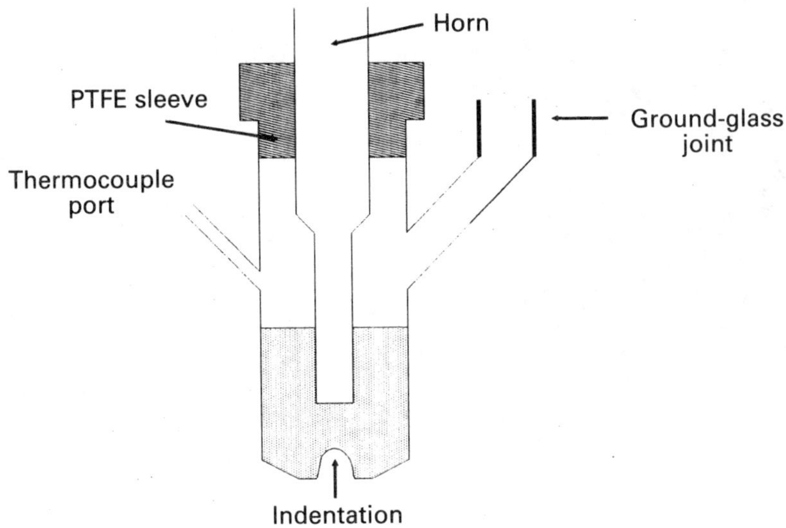

Fig. 3.6 — Dimple cell.

(c) Rosett cell

The Rosett cell was originally designed to be operated 'open-ended' for biological cell disruption. It allows the irradiated reaction mixture to be sonically propelled from the end of the probe around the loops at the base of the vessel, thus providing both cooling (when the vessel is immersed in a thermostatted bath) and efficient mixing. When equipped with a flanged lid and a PTFE seal this is an excellent general purpose sonochemical reactor (Fig. 3.7).

(d) Pressure cells

In situations where elevated pressures need to be used a useful reactor has been described by Suslick [3]. Here the screw thread on the horn (see section 3.3.2) is used

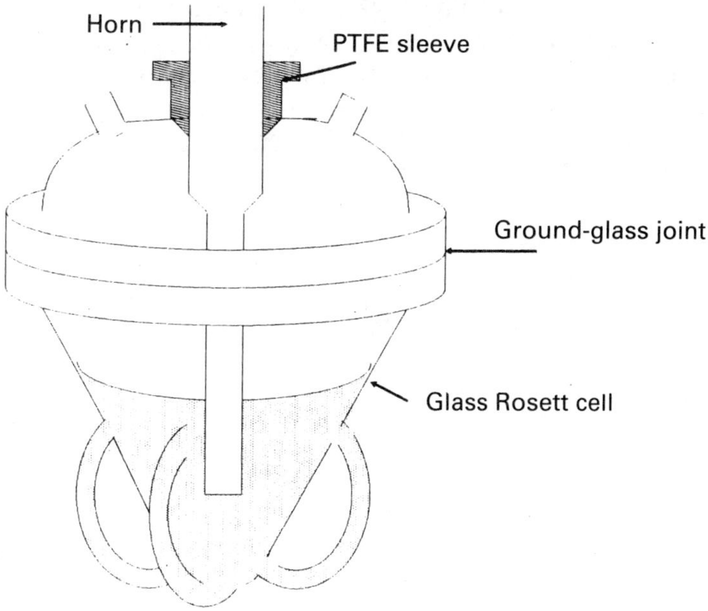

Fig. 3.7 — Rosett cell.

to attach a PTFE collar which locks onto the outside of the glass joint of the sonochemical apparatus (Fig. 3.8). This provides a much more efficient pressure seal than the normal push-fit of the PTFE sleeve into ground-glass joints described above (Fig. 3.5).

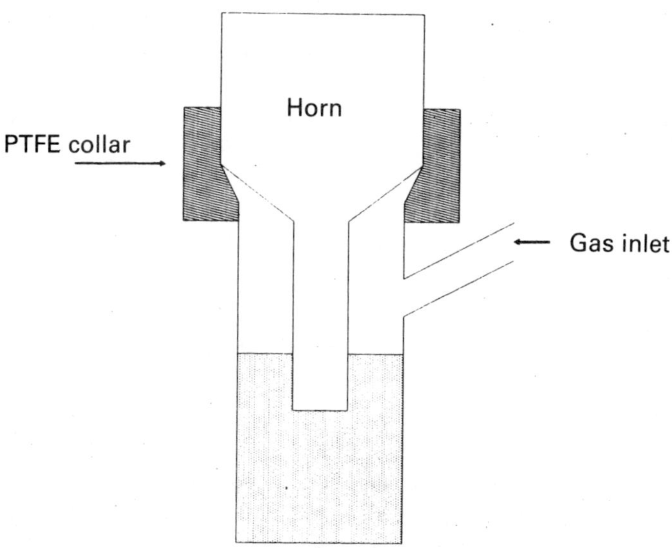

Fig. 3.8 — Simple pressure cell.

A more refined solution has been developed by Undatim Ultrasonics for their variable frequency Sonoreactor. This cell provides an adjustable anvil to give the optimum reflection and therefore a better coupling of the ultrasonic power to the liquid system (Fig. 3.9).

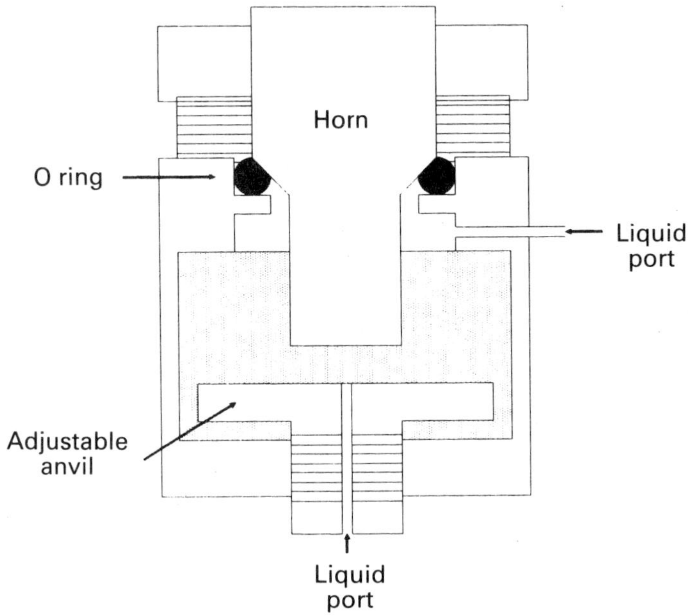

Fig. 3.9 — Pressure cell with adjustable anvil.

It must be borne in mind that, under pressure, the cavitational collapse is much more violent so that, in addition to a possible greater sonochemical effect, there will also be greater erosion of the probe tips used in the reactors.

A limitation of the reactors described above is that they are all of the batch type. A favourite method of avoiding this is to use a flow cell in a continuous or circulating system (see Chapter 4).

(e) Cooled cell with reagent holder

A piece of apparatus developed in the laboratories of Luche is shown in Fig. 3.10 [4]. External cooling is provided by a water jacket, and the small glass 'tray' is used to hold a metal reagent close to the horn tip, i.e. permanently in the maximum cavitation zone. This cell was originally used for the generation of dialkyl zincs. In Chapter 5 an experiment utilizing this cell is described.

3.4 THE EFFECT OF EXTERNAL PARAMETERS ON SONOCHEMISTRY

There are a number of physical parameters which can be changed in order to affect the action of power ultrasound. In this section we report the results of an investigation into how ultrasonic energy input to a chemical system is affected by the following important reaction parameters: ultrasonic power used, the presence of

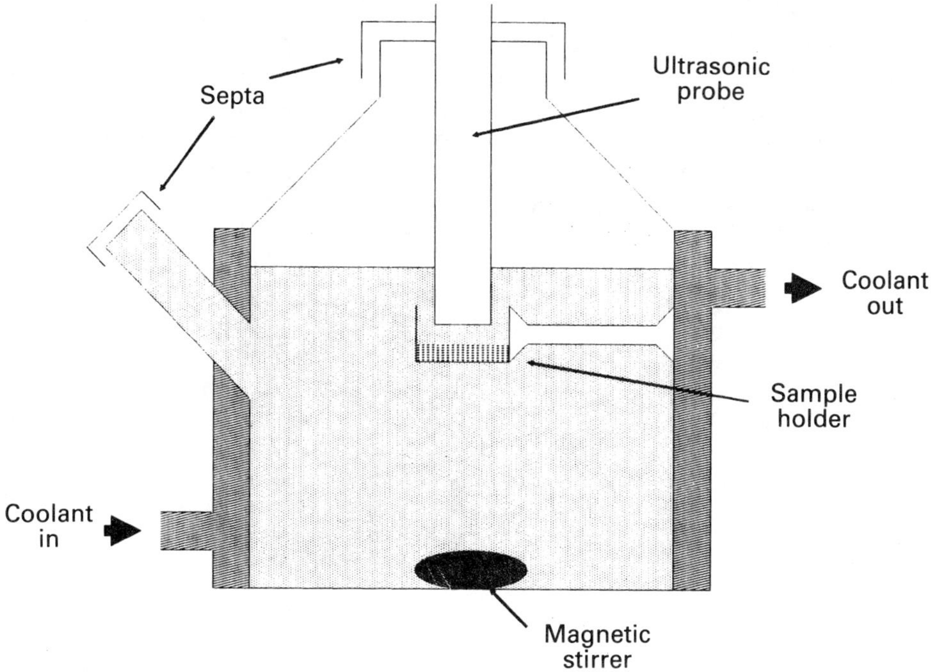

Fig. 3.10 — Cooled cell with reagent holder.

bubbled gas, temperature, solvent composition, solid–liquid ratio and reaction volume. The work was performed using the Undatim Sonoreactor probe system, which is equipped with an automatic transducer resonance frequency search device, enabling the power input to a system to be maintained accurately throughout a reaction, thereby offering considerable advantages over the normal commercially available equipment in terms of both the monitoring and control of irradiation parameters.

3.4.1 The system for study

Of the many different classes of reaction which can be enhanced by ultrasonic irradiation the most common are heterogeneous, including both liquid/liquid and liquid/solid. The chosen system for study was the particle size reduction of copper bronze in liquid media under a variety of different conditions [5]. Although this is not a 'chemical' reaction as such, the reduction of particle size, i.e. increase in available surface area of a reactant or catalyst, is a major contributor to sonochemical enhancement of solid/liquid reactions. Moreover, the size reduction is a readily quantifiable factor and thus ideal as a guide to estimating sonochemical efficiency for the range of conditions chosen.

It is important to note that the results obtained in this study are based upon a solid/liquid heterogeneous model reaction and so may not be directly applicable to heterogeneous liquid/liquid or homogeneous reactions, for which different models may be necessary.

3.4.2 The effect of bubbled gases

If the main cause of sonochemical effects is cavitational collapse then it is clear that sonochemical efficiency must be linked directly to the generation of cavitation bubbles in the chemical reaction. For reproducible results many workers have used a bubbled gas to generate large numbers of nucleation sites for cavitation and provide bubbles of uniform 'energy' of collapse. Most effective for such processes are the mono-atomic gases such as argon, with diatomics such as nitrogen next and polyatomics a very poor third. This is directly related to the ratio of specific heats for these gases, the highest ratios giving the greatest cavitational effects. The particle size reduction of copper bronze, induced by sonication, illustrates the effect of entrained gas on sonochemistry (Table 3.1). The results clearly show that, if all other external parameters are constant, there is a distinct advantage in the continuous passage of argon through the system.

Table 3.1 — The effect of bubbled gas and temperature on the ultrasonically induced particle size reduction of copper bronze in water

Gas used		Particle size (after 1 h) (μ)
No gas		23.8
Nitrogen		22.9
Argon		19.1
Temperature (°C)	Intensity ($W\,cm^{-2}$)	
90	23	31.9
60	39	29.3
30	62	23.8
0	79	20.6

Copper bronze original particle size 60 μm (2.0 g) in water (100 cm^3), intensity 65 $W\,cm^{-2}$ at 30°C, 20 kHz.

The only disadvantage of such methodology in general sonochemistry is that, at reaction temperatures above ambient, there could well be losses of volatile material through 'evaporation'. It has been demonstrated that efficient saturation of a gas with liquid vapour is achieved by insonation of that liquid with the gas passing through [6]. For this reason in the studies reported here the insonation has been performed in the absence of entrained gases. Except where stated, all reactions were performed using a 'dimple cell'.

3.4.3 The effect of reaction temperature

As the temperature of a reaction is increased, the vapour pressure of any liquid involved in the reaction is also increased. Since the collapse of a cavitation bubble, formed during sonication of the medium, will be directly influenced by the quantity of vapour which enters the bubble during its formation, (i.e. the more vapour which enters the bubble the more cushioned is the collapse), any increase in the temperature (i.e. vapour pressure) of a system will influence cavitational bubble collapse by effectively increasing the collapse time. Thus as the system temperature is raised, the ultrasonic power input will need to be increased if constant cavitational effect — approximating to a uniform sonochemical effect — is to be maintained. Alternatively, if the ultrasonic power setting on the generator is kept constant then the lower the temperature of operation the greater will be the sonochemical effect. This is demonstrated in Table 3.1, where the Undatim Sonoreactor, operating at 20 kHz, was set at 7 during the insonation of water at four different temperatures.

The effect of reducing the bulk water temperature from 90°C to 0°C, in steps of 30°C, on the reduction in particle size of copper bronze powder (Table 3.1), shows clearly the improvement achieved as the temperature is lowered.

3.4.4 The effect of reaction temperature and solvent composition on irradiation frequency and power

If cavitation is the major cause of sonochemical effects then factors such as the vapour pressure, viscosity and surface tension of the solvent will have a bearing on sonochemical reactions. Any change in the reaction temperature results in a change in these parameters with, as a consequence, a change in the optimum resonance frequency for the system. For example, employing the frequency search mode of the Undatim instrument during the irradiation of 100 cm^3 of water, at different temperatures and employing three different transducers, a small but significant change in frequency was observed (Table 3.2). The precise change depends upon the geometry of the vessel and the position of the probe tip and is much more pronounced in heterogeneous systems.

Table 3.2 — Optimum resonance frequencies at different temperatures using transducers of differing nominal frequencies

Nominal frequency (Hz)	Temperature (°C)	Resonance frequency (Hz)
20 000	30	20 668
20 000	60	20 627
20 000	90	20 578
40 000	30	39 262
40 000	60	39 128
40 000	90	39 024
60 000	25	59 652
60 000	50	59 486

These studies performed using 100 cm^3 water in a glass beaker.

All chemical reactions, by definition, proceed to a product which differs from the starting material, i.e. the properties of the reaction mixture change with time, therefore there will be a change in the acoustic properties of the medium during the course of a reaction. As a model for this change, 200 cm^3 of water and dimethylformamide (DMF) at 20 kHz were separately sonicated and the power input to each system was monitored. At instrument setting 7 it proved possible to transmit 122 W of power to the aqueous system but in DMF the power dissipated at the same setting was only 101 W.

Thus, in order to maintain optimum performance for any sonication system where the reaction temperature is not constant, it may be necessary, with conventional instrumentation, to adjust the tuning of the transducer during the course of the reaction.

3.4.5 The effect of changes in the insonation frequency on chemical reactivity

One of the problems associated with operating at different insonation frequencies is that the physical dimensions of the transducers and horns place a real limit upon the power (i.e. amplitude of vibration) which they can deliver. A comparison of the amount of energy (calorimetric) dissipated in 100 cm^3 of water at 30°C by three different horn systems is shown in Table 3.3. In the same table the physical consequences of this can also be illustrated experimentally by observing the rate of particle size reduction for copper bronze at 30°C.

Table 3.3 — The effect of changes in irradiation frequency on optimum power delivered and sonochemical efficiency

Nominal frequency (kHz)	Power setting	Tip area (cm^2)	Power (W)	Intensity ($W\,cm^{-2}$)	Particle size (μm)
20	7	1.40	96	68.6	23.8
40	5	1.00	26	26.0	37.7
60	4	0.75	11	14.7	43.2

Copper bronze original particle size 60 μm (2.0 g) in water (100 cm^3), sonication time 1 h, 30°C.

It should not be concluded from these results that the optimum sonochemical effect can only be obtained at high powers, i.e. at low (20-kHz) frequency (see sections 1.4.1 and 1.4.8). From these results it is certainly tempting to assume that the greatest effects are achieved at lower frequencies. What must be borne in mind, however, is that the actual ultrasonic power entering the system also depends upon the frequency — high frequencies are delivered at lower powers. Thus in order to properly equate the effects of ultrasonic irradiation at different frequencies the time of exposure must be considered. A report on the deagglomeration and dispersion of

SiC has effectively made this point [7]. In order to deagglomerate a material with ultrasonic irradiation two main parameters must be considered:

(a) A threshold power level will be necessary to obtain any size reduction and this may be dependent on the frequency.
(b) Whatever frequency is used, it is the total sum of ultrasonic energy delivered to the system which is important.

Thus one might expect that the degree of deagglomeration will be dependant on the combination 'time × energy' and results should be more meaningful if the data is compared using this measure of the total energy delivered.

Using the Undatim system the effect of producing the same total energy from a system using 20-, 40- and 60-kHz probes can be seen in Table 3.4. The results clearly indicate that, even if the same total energy is used, the 20-kHz probe system does appear to give a better size reduction than the 40- or 60-kHz systems. Significantly, however, both the 40- and 60-kHz systems give very similar results. The same experimental comparison was also made using two different powders, aluminium oxide and nickel, both suspended in water. Again, the same trend was seen. The 20-kHz probe produced a greater particle size reduction than the 40- or 60-kHz probe at the same energy delivered.

Table 3.4 — Particle size reduction achieved at a total power input of 480 W min

Frequency (kHz)	Energy (W)	Time (min)	Particle size (μm)		
			Cu	Ni	Al_2O_3
20	96	5.0	36.44	51.00	64.50
40	26	18.5	43.46	56.64	70.96
60	11	43.6	44.17	57.96	74.01

Temperature 30°C, 100 cm^3 water in a dimple cell. Original particle size Cu 60μ; Ni 70μ; Al_2O_3 85μ.

The effect of insonation frequency in sonochemistry is, however, a relatively new area of study, and indications are that in homogeneous reactions and those involving emulsification there may well be a significant frequency effect. This may well be related to the physics of such systems rather than their chemical reactivity.

3.4.6 The effect of reaction volume

Intuitively one might expect that for any given input power the sonochemical effect would diminish as the volume of the reaction mixture was increased. The results obtained from the Undatim equipment (Table 3.5) are therefore somewhat surprising. The results suggest that great care is required when calculating the power input required to scale up a sonochemical reaction.

Table 3.5 — Power generated (20 kHz) in various volumes of water

Volume (cm^3)	Power delivered (W)	Intensity ($W\,cm^{-2}$)
50	84	60
100	95	68
200	120	86

Area of Tip 1.4 cm^2, power setting 7.

3.5 THE ADVANTAGES AND DISADVANTAGES OF USING AN ULTRASONIC PROBE SYSTEM FOR SONOCHEMISTRY

In the chemical laboratory the commercially available probe system is the best method of introducing high power ultrasound into a chemical reaction. Its advantages and disadvantages are summarised below.

3.5.1 The advantages

(a) In a given medium the ultrasonic power delivered by a horn is directly related to the magnitude of vibration of the tip. This can be readily controlled by the power input to the transducer, and so the precise power of the system can be regulated. Input powers of over 100 $W\,cm^{-2}$ can be easily achieved (depending on the size of the unit).

(b) Ultrasonic streaming from the tip of the probe operated at moderate power is often sufficient to provide bulk mixing.

(c) Most modern units have a pulse facility, allowing the operator to sonicate reactions repeatedly for fractions of a second. This gives adequate time for 'cooling' between sonic pulses.

(d) With such systems the probe system can be tuned to give optimum performance. This is important in order to obtain reproducibility of results but it is also important to use carefully controlled procedures in terms of reaction vessel shape and size.

3.5.2 The disadvantages

(a) As with all systems which operate using piezoelectric transducers the optimum performance is obtained at a fixed frequency. For most commercial probe systems this is 20-kHz and, although it is possible to drive them at their overtones (i.e. 40 or 80-kHz), the power dissipation at overtones of the fundamental frequency of the system is very much reduced. To operate successfully at different frequencies it is best to purchase individual amplifier/horn systems tailored to individual requirements.

(b) As with baths there is a problem over accurate temperature control unless precautions are taken. The use of specially designed reaction vessels, e.g. a

Rosett cell, does alleviate much of this difficulty. The temperature within the reaction vessel should be continuously monitored.

(c) Owing to the high intensity of irradiation in the reaction close to the tip, it is possible that radical species may be produced which could interfere with the normal course of reaction.

(d) The cavitation which is the source of chemical activation is also the source of a common problem with probe systems — tip erosion — which occurs despite the fact that most probes are fabricated of titanium alloy, which is very resistant to erosion. There are two unwanted side-effects associated with erosion: (a) metal particles eroded from the tip will contaminate the reaction mixture and (b) the physical shortening of the horn causes a loss of efficiency (eventually it will become too short to be tuned). The latter problem is avoided by the use of screw-on tips to the probe in the form of studs; this eliminates the need for a costly replacement of the whole horn.

(e) Special seals will be required if the horn is to be used in reactions involving reflux, inert atmospheres or pressures above (or below) ambient.

(f) Generally they are only suitable for small batch reactions, although multiple probes will cope with larger volumes.

REFERENCES

[1] Sonic Systems, Monks Dairy, Isle Brewers, Taunton, Somerset TA3 6QL, UK; Undatim Ultrasonics, Parc Scientifique, 10 Rue du Bosquet, B 1348 Louvain la Neuve, Belgium.

[2] Branson Ultrasonics, Eagle Road, Danbury, Connecticut 06810-1961, USA; Heat Systems, 1938 New Highway, Farmingdale, NY 11735, USA; Sonics and Materials, Kenosia Avenue, Danbury, Connecticut 06810, USA.

[3] K. S. Suslick and S. J. Doktycz, Advances in Sonochemistry, ed T. J. Mason, Vol 1, 197, JAI Press, 1990.

[4] J. C. de Souza Barboza, C. Petrier and J.-L. Luche, *J. Org. Chem.* (1988), **53**, 1212.

[5] T. J. Mason, J. P. Lorimer and D. M. Bates, Ultrasonics (1991) **29**, in press.

[6] T. J. Mason and P. Sephton, *Aldrichimica Acta* (1989) **22**, 2.

[7] M. Aoki, T. A. Ring and J. S. Haggerty, *Advanced Ceramic Materials* (1987) **2(3A)**, 209.

4

Scale-up and flow systems

It has been recognized for many years that power ultrasound has great potential for uses in a wide variety of processes in chemical and allied industries. Reported applications include cleaning, sterilization, flotation, drying, degassing, defoaming, soldering, plastic welding, drilling, filtration, homogenization, emulsification, dissolution, deaggregation of powder, biological cell disruption, extraction and crystallization.

The interest of chemists in power ultrasound has also been established whether it be to improve yields, speed up chemical or physico chemical processes or, as seems possible in some cases, completely switch reaction pathways. And yet, in spite of the great amount of information available on sonochemistry, and in spite of the very promising effects described on a laboratory scale, the literature does not seem to provide many details of industrial applications. Over the last few years a number of groups have become interested in solving the problems of scale-up, and the purpose of this chapter is to give an overview of the available systems that can be used to carry out sonochemical transformations on a scale rather larger than that used in the research laboratory [1].

4.1 THE CHALLENGE OF SCALE-UP

There are essentially two types of large-scale chemical plant: batch and flow type. Sometimes the flow system will form a part of batch processing as a loop attached to the main vat. In the preceeding chapters we have concentrated on laboratory-scale equipment. Some of these small-scale systems can be readily adapted for scale-up, but a precise knowledge of the type of sonochemistry involved in the process is essential. The first decision to be made when considering whether a sonochemical reaction should be scaled up is whether that reaction is in fact truly sonochemical or simply the result of an ultrasonically induced mechanical effect. In a solid/liquid heterogeneous reaction, for example, power ultrasound may only serve as an efficient dispersing system (for particle fragmentation/deagglomeration and dispersion). In this situation an ultrasonic pretreatment of a slurry may be all that is required before reaction is allowed to take place conventionally. However, if ultrasound has a real effect on the chemistry of the system then arrangements must be made to provide sonication during the reaction itself.

To scale up a true sonochemical process, there are three main questions to be answered:

(a) What will be the most appropriate sonolysis system for the reaction mixture to be processed?
(b) What are the best reaction conditions?
(c) What will be the energy implications of using ultrasound rather than traditional methodology?

These questions are not easily answered because of the number of factors which must be taken into account when judgement is made. Thus to attain optimum conditions for sonication the variables which can influence cavitation must be carefully studied. These variables depend upon:

(i) The reaction medium itself: in terms of its viscosity, vapour pressure, the nature and concentration of any dissolved gas and the presence of any solid particles before, during or after reaction.
(ii) The reaction conditions: including the temperatures and pressures involved — these may well be varied during a conventional process.
(iii) The type of sonic system used: this must involve not only the power required but also the best frequency to be used and the size and geometry of the chemical reactor to be employed.

4.2 SONOCHEMICAL EFFICIENCY

Power ultrasound can be introduced into a system by either (a) dipping a vessel containing the reaction mixture into a tank containing a sonicated liquid (most generally water), (b) using a reaction vessel whose walls are subjected to ultrasonic vibrations, or (c) immersing a source of power ultrasound into the reaction medium itself. Whichever method is employed, the first place where any loss in efficiency can arise is in the design of the generator where mains power is converted into a signal to the transducer. At this stage there is an inevitable loss of heat. The motion of the transducer will itself involve further losses of energy in the form of heat and sound. Further inefficiency arises from the coupling of the transducer to its load, i.e. the mass which it is insonating. Lastly there is the problem of whether it is in fact possible to achieve sonication throughout the whole reaction mixture or whether it would be easier to localize the power ultrasound in a particular region of the reacting system.

Scheme 4.1 can be considered to represent a sonochemical process from the energetics viewpoint. The ultrasonic system transforms electrical power into vibrational energy, i.e. mechanical energy. This mechanical energy is then transmitted into the sonicated reaction medium. Part of it is lost to produce heat, and another part produces cavitation, but not all of the cavitational energy produces chemical and

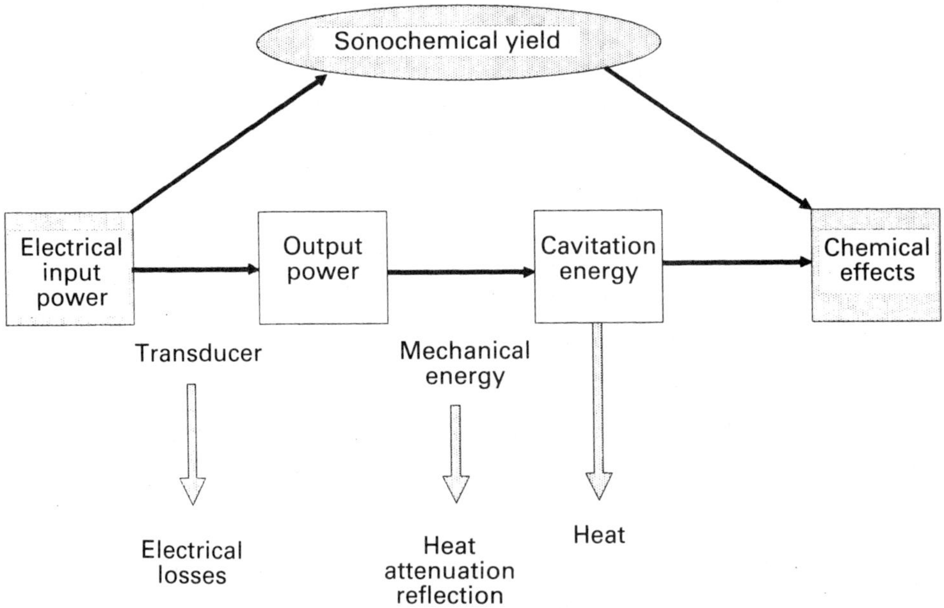

Scheme 4.1

physical effects. Some energy is reflected and some is consumed in sound re-emission (harmonics and sub-harmonics). From this, it can be appreciated that the determination of an energy balance is not easy. Nevertheless one can define the term 'sonochemical yield' (S.Y.), as:

$$\text{S.Y.} = \text{measured effect/input power in watts}$$

This is similar to the definition of 'photochemical yield'. The measured effect can be, for example, the number of moles of product generated per second, and the input power can be monitored and measured with a wattmeter.

To estimate the output power, i.e. the actual power entering the system, one has to know the conversion yield of the ultrasonic system used. This is not always obvious, since part of the energy is reflected back to the transducer, especially when the sonicated mixture is viscous or contains a large amount of suspended solid. Several attempts have been made to measure the overall output power (see Chapter 1). In a pure liquid, one might assume that almost all the mechanical energy produces heat and so, via calorimetry, some estimate of the output power can be obtained. However, care should be taken in circumstances where sonochemical degradation of the liquid occurs. For example, sonicated water produces hydrogen and hydrogen peroxide, which is an endothermic reaction ($\Delta H = 383$ kJ mol^{-1}). If this reaction occurs in, say, a 1% yield, the absorbed energy is of the same order of magnitude as an increase in the temperature of 1 l of water by 1°C.

Other attempts have been made to monitor output power using model sonochemical reactions [21]. One such reaction is the production of iodine from a sonicated aqueous solution of potassium iodide. This is not necessarily an accurate measure of output power; it is only a measure of the overall sonochemical yield of the whole reaction under the particular sonochemical conditions employed. At the time of writing there is still no accurate method for measuring the overall 'sonic energy', and this therefore remains as a major area of research in scale-up.

Another tricky problem is the determination of the way in which the 'sonic energy' is distributed within a particular reactor configuration using a given sonic system. One approach developed for the determination of this distribution involves thermistor probes [2]. Pugin has shown that the distribution of ultrasonic energy is not at all homogeneous [3]. This has been accomplished in a very simple way, by measuring the local temperature at different places in the sonicated media. Maxima and minima appear, depending on the shape of the vessel, the liquid height, the input power and the nature of the sonic system (bath or probe). In some cases standing waves may be formed, and there is a question as to whether a larger sonochemical effect is obtained with standing waves or in their absence [4].

The energy distribution might well have some effect on the course of a chemical reaction, and some evidence for this has been given recently in the case of a solid–liquid phase transfer reaction (Eq. 4.1) [5]. In this Michael addition of the anion of pentan-2,4-dione (2) to chalcone (1), the adduct (3) is formed and this can react further to give the cycloadduct (4). The overall *rate* of this reaction is increased by sonication, but the chemical *selectivity* depends on the source of irradiation.

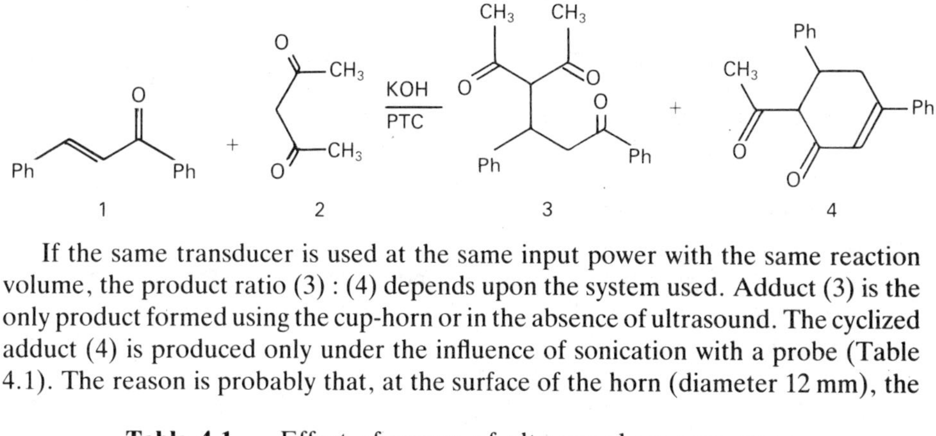

1 2 3 4

If the same transducer is used at the same input power with the same reaction volume, the product ratio (3) : (4) depends upon the system used. Adduct (3) is the only product formed using the cup-horn or in the absence of ultrasound. The cyclized adduct (4) is produced only under the influence of sonication with a probe (Table 4.1). The reason is probably that, at the surface of the horn (diameter 12 mm), the

Table 4.1 — Effect of source of ultrasound on percentage product distribution in a condensation reaction

No sonication		Cup-Horn		Probe	
(3)	(4)	(3)	(4)	(3)	(4)
52	0	69	0	72	12

10-min reaction time, room temperature, 1 W cm^{-3}.

'power density' (expressed in Watts per square centimetre) is higher than at the surface of the cup-horn (diameter 45 mm). This probably results in more severe conditions in a small volume directly under the horn, which can promote the cyclization side-reaction.

This difference in product distribution points to one obvious problem when scaling up sonochemical processes. In an ideal situation one would establish a model reaction on a laboratory scale and assess the input power required for the optimum sonochemical yield. The energy input could then be characterized as the energy distribution in the reactor either as the energy emitted at the surface of the ultrasonic device (expressed as Watts per square centimetre) or the energy dissipated in the bulk of the sonicated medium (expressed as Watts per cubic centimetre). In the scaled-up version of the reactor the shape and type of device may not be directly comparable to the laboratory reactor, and so product differences may arise. However, this problem of comparable energy distribution is not likely to be so important when dealing with the smaller sonicated volumes such as those used with small continuous-flow cells or the RUM system.

4.3 LARGE-SCALE SONICATOR SYSTEMS

Solutions to the problem of the scale-up of sonochemical reactions do exist, but they are not as simple as the use of bigger versions of laboratory equipment. In a production situation the volumes treated will be very much larger than those considered in the laboratory, and the type of process will govern the choice of reactor design. It could well be that some processes would be more suited to low intensity sonication (e.g. using a bath type reactor or a liquid whistle) whereas others may need higher intensity irradiation (via a probe-type system).

At the end of the 1970s there were only a limited number of pieces of larger-scale equipment available for sonochemistry — the whistle and the bath. Liquid whistles were, and are, used in industrial plants for homogenization and emulsification, capable of processing up to several cubic metres per hour [6]. Gas whistles have also been used for drying. There is no doubt that these are very economic systems from an energetic point of view and have many other potential applications, such as liquid–liquid extraction and liquid–liquid phase transfer reactions. The restriction with this type of equipment is that attempts to process solid–liquid reaction mixtures can result in rapid erosion of the vibrating blade. Large-scale cleaning baths have been available for many years with capacities up to several cubic metres. Some of the more rugged types, capable of prolonged use without maintenance, may well employ the older type of magnetostrictive (rather than piezoelectric) transducers.

Quite large-scale continuous-flow-cells were also available before 1970 [6], although at that time they did not seem to have found applications on an industrial scale. The reason why very few scale-up attempts have been reported is probably because none of these systems appear to be convenient for chemical applications. The situation has rapidly changed, however, over the last five years. This is due to the increasing and widespread interest in sonochemistry on a laboratory scale. Some ultrasonic equipment manufacturing companies have started looking again at devices already in their inventories which might usefully be adapted and employed by chemists, while others have come up with new and original devices better fitted to chemical or physical transformations.

Until recently a question mark hung over whether large-scale sonochemistry equipment could be devised for industrial application. The problem was not simply that large volumes needed to be treated but also that studies of the effect of changes in reaction parameters suggested that carefully optimized reaction conditions might be required. Perhaps 'tailor-made' reactors would be needed to achieve optimum sonochemical yields. Fortunately, however, although the effects of some variables, including the nature and the shape of the sonoreactor must be very well defined, there can be little doubt that the required equipment can easily be derived from the wide-ranging types of systems which are listed in Table 4.2.

Table 4.2 — Types of sonicator system commercially available

1	Cleaning bath (including submersible transducers)
2	Probe system (including the flow cell, cup horn)
3	Shaking tray
4	Liquid whistle
5	Tube reactor (polygon or pipe type)
6	RUM
7	Push–pull

4.3.1 Cleaning bath systems

(a) The ultrasonic cleaning bath

In cases where only low intensity ultrasonic irradiation is needed, batch treatment could well be as simple as immersing the reaction vessel in a large-scale ultrasonic cleaning bath. This was the method adopted by Repic in his cyclopropanation reactions (see section 5.1.3). If the cleaning bath itself is to be used as the reaction vessel the bath tank itself would need to be constructed of a material which was inert towards the chemicals involved. An appropriate grade of stainless steel might prove adequate, or plastic tanks could be used. In the latter case, however, the transfer of ultrasonic energy from the tank walls to the reaction mixture would be attenuated by the plastic, and so the transducers would need to be bonded onto a stainless steel or titanium plate and this assembly then bolted to the inside wall of the tank. In addition the tank could well require a sealed lid with addition ports, and almost certainly stirring would be required.

(b) Reaction vessel with external transducers

A more sensible solution is to adapt an existing reaction vessel by the addition of external transducers (Fig. 4.1). This arrangement has the advantage that facilities for

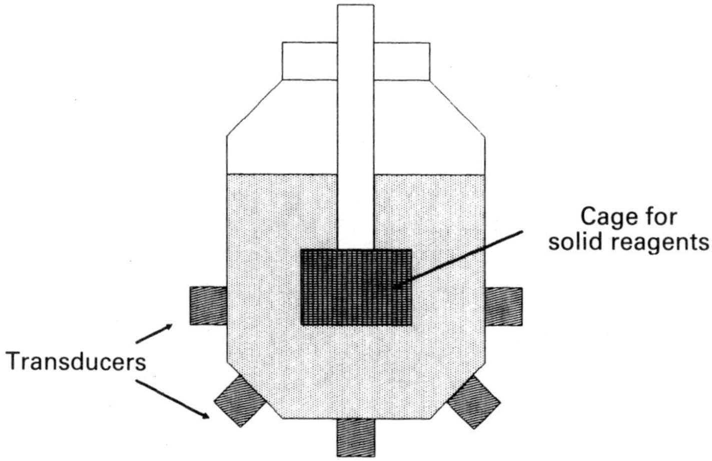

Fig. 4.1 — 'Cleaning bath' reactor.

stirring and for the use of elevated temperatures and pressures would already be available. There are, however, some disadvantages:

(i) The acoustic energy entering the reaction will be quite small, and the presence of a stirrer or solids will cause attenuation of the sound energy. This may lead to a central zone which is not under sonication, thus placing a limit on the maximum size of the vessel. In this configuration the energy distribution will also be hard to determine.
(ii) If the surfaces to which the transducers are to be attached are curved then very careful design will be required to attain the best energy transfer.
(iii) For heated reactions the transducers will need to be cooled and/or removed from direct contact with the vessel walls. The use of wave guides will enable the transducers to be positioned away from the vessel walls (at some multiple of the half wavelength of the sound in the wave guide material, i.e. the wave guide is effectively a horn extender as described in Chapter 3).

Bath-type reactors can also be used in a flow system. In this case the reacting liquids could be continuously fed into an ultrasonic tank with outflow to the next process. Such treatment could be intensified by recycling or by connecting a number of such sonicated tanks in line.

(c) The submersible transducer
A useful variant to the sonicated tank is the submersible transducer assembly. In fact this offers somewhat greater flexibility in use since it comes as a sealed unit capable of withstanding organic solvents — this dates from the days when halocarbon-type cleaning fluids were in common usage (see Fig. 2.8). Such units can be designed to fit

into any existing reaction vessel. As with the cleaning bath, ultrasonic reactors involving submersible transducers would require some form of additional (mechanical) stirring.

4.3.2 The probe system

(a) Batch Reactors

The ultrasonic probe system, with the various types of horn design available to amplify vibrational amplitude (see Chapter 3), is the most effective method of introducing high power ultrasound into a liquid. Despite this there are some severe restrictions in usage for large-scale batch treatment:

(i) The high energy density produced from a single probe tip is unlikely to be capable of delivering sufficient energy density to affect the whole of the reacting volume, despite the circulation induced by streaming. Unlike the bath system described in section 4.2.1(b) above, a probe reactor tends to have its 'dead zones' on the periphery of the vessel (Fig. 4.2).

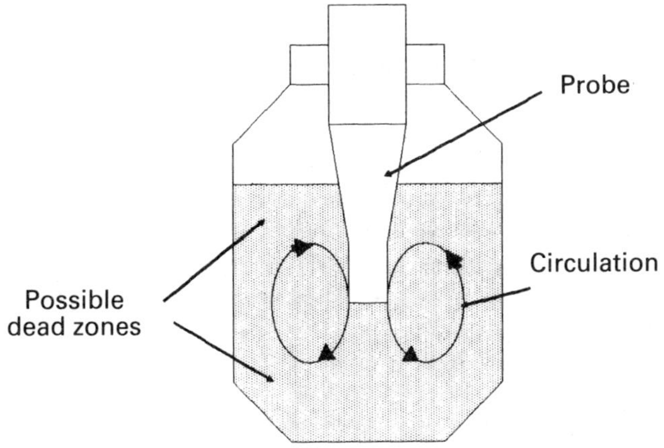

Fig. 4.2 — Probe batch reactor.

(ii) In some reactions an optimum power level can be reached (which depends on the system) beyond which increased power input actually reduces the yield.

(iii) At high powers, erosion of the probe tip will contaminate the reaction and will also necessitate frequent probe replacement.

(b) Flow System

For larger-scale operations where high intensity ultrasound is required the sonic horn is best utilized as part of a flow loop outside of the main reactor (Fig. 4.3). This permits the processing of large volumes and provides several major advantages over the batch configuration for the probe system:

(i) High intensity sonication is provided for a continuously flowing liquid reaction at controllable powers and with a pulse option.

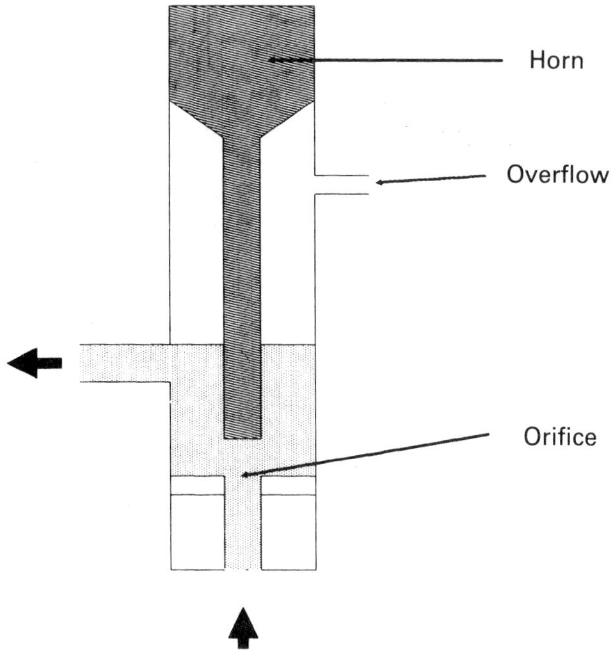

Fig. 4.3 — Flow cell.

(ii) Temperature control is provided through the circulating reaction mixture.
(iii) An additional control feature is possible for this system by varying the residence time of the reacting system in the flow cell, i.e. flow rate.

One might, therefore, expect that flow systems would be favourite for high intensity treatment of large volumes. The simple flow cell is an excellent means of processing relatively large volumes. For plant use, however, it is likely that a design modification would be needed which involves the coupling of a probe transducer into a flow pipe by means of 'T' sections (Fig. 4.4). A number of such transducers could be employed in this manner to give extended treatment time.

There are certain disadvantages to the flow system:

(i) High power treatment will produce tip erosion and may generate radicals.
(ii) Pumping is required for circulation to the flow cell and so it is not suitable for either very viscous reaction media or heavily particulate systems.

4.3.3 The vibrating tray

This device was originally designed by the Lewis Corporation [7] for the processing of coal or metal ores at rates up to 20 tons/h (Fig. 4.5). The block of transducers is bonded to the lower face of the tray and the vibrations are transmitted along its length. The tray assembly must be suspended on elastic supports so that the

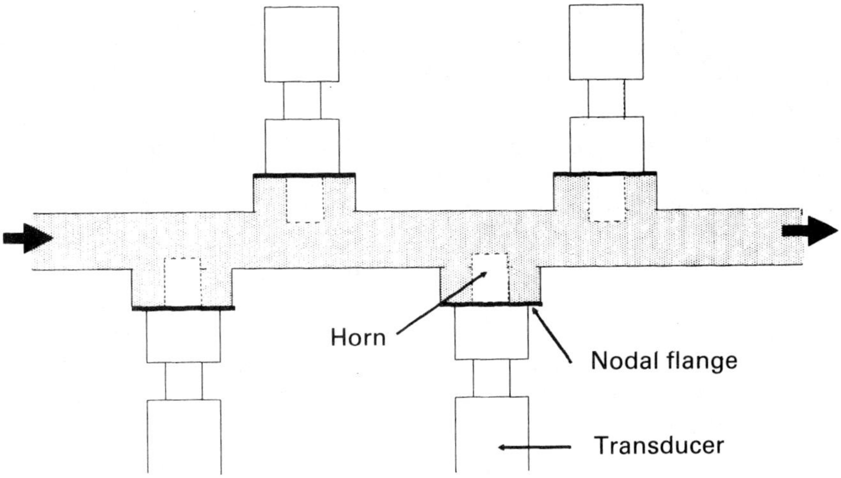

Fig. 4.4 — Standard pipe with T-piece inserts.

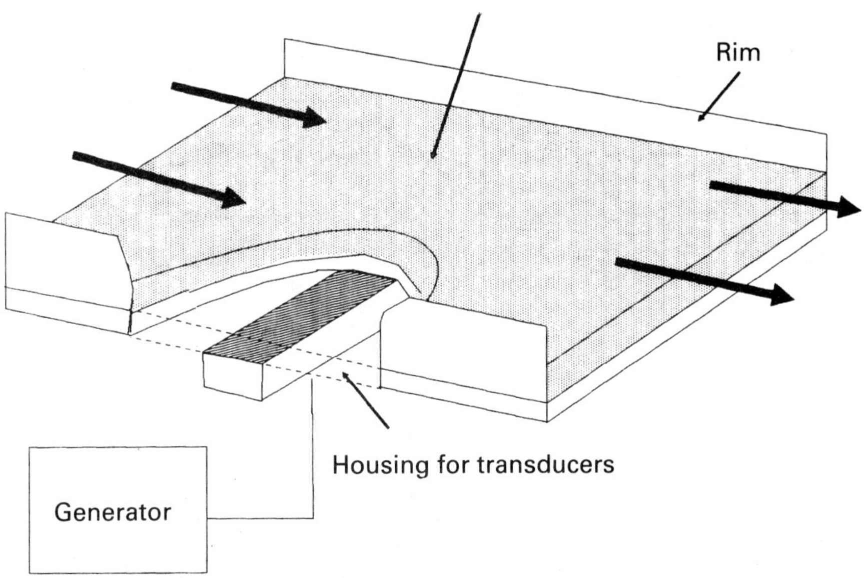

Fig. 4.5 — Vibrating tray.

ultrasonic vibrations are not absorbed through solid mountings. In the particular case of coal recovery from waste tips the process involves mixing the coal 'waste' with equal quantities of water and, after crude screening to remove rocks, allowing gravity to carry the mixture down the vibrating tray (20 kHz). With a residence time of only a few seconds this process yields marketable low ash coal product with clay and sand suspended in the wash water, which can be allowed to settle in a pond so the water can be recycled for further processing.

The vibrating tray is generally used in open-air conditions and is clearly both large-scale and extremely robust. Design modification to allow the unit to be installed within a pressure vessel would permit the processing to be used at elevated temperatures with complete containment of vapours. This then becomes a piece of heavy duty chemical processing equipment for solid/liquid reactions or extractions.

4.3.4 The liquid whistle reactor

The liquid whistle was originally introduced as an efficient mixing/homogenization system and this still remains one of its main applications. As the name suggests it operates on the whistle principle in that ultrasonic vibrations are generated in a steel plate as a liquid is passed across it at high velocity. The liquid itself couples with the vibrations to produce cavitation in the flow, which, in turn, leads to extremely efficient mixing (see Fig. 1.8) [8]. In many chemical applications such mixing/ emulsification/homogenization is an essential requirement for promoting reactivity.

These devices differ markedly from the more usual bath and probe types in that they derive their power from the medium (by mechanical flow across the blade) rather than by the transfer of energy from an external source to the medium. The majority of the chemical effects observed on using whistle-type transducers for the sonication of heterogeneous reactions can be attributed to the generation of very fine emulsions rather than the ultrasonic irradiation itself.

The use of the liquid whistle type of ultrasonic generators for homogenization has increased dramatically since the Second World War. It was as far back as 1927 when Wood and Loomis reported that oil and water could be emulsified on sonication in the same beaker using quartz piezoelectric transducers [9]. However, at that stage in the development of ultrasonic equipment it was quite difficult to see how transducers driven by quartz crystals could be employed on an industrial scale, and so ultrasonic homogenization remained a curiosity until 1948. In that year Janovski and Pohlmann highlighted the economic advantages to be gained from the use of a liquid whistle compared with magnetostrictive transducers of the type available at that time [10]. In 1960 a series of experiments was undertaken to compare four methods then in common usage for the emulsification of mineral oil, peanut oil and safflower oil [11]. The results proved that a homogenizer which operated via a liquid whistle was superior to three other types of apparatus, namely a colloidal mill and two types of sonicator, one of which employed a quartz crystal and the other a barium titanate transducer. Subsequently the liquid whistle was adopted as an ideal method of homogenization, and it is this same design of instrument which has been used by Davidson to enhance the hydrolysis of fats and waxes (Fig. 1.8, see also section 5.3.2) [12].

The method has several advantages over flow systems involving probes:

(i) The system was developed, and is ideal for, such processes as emulsification, homogenization and dispersion.
(ii) It can be used for the processing of flow systems and, as such, has immediate possibilities for scale-up.
(iii) With no moving parts, other than a pump, the system is rugged and durable.

An obvious benefit of such a system is that it can be used for flow processing and can be installed 'on-line'. In this way large volumes can be processed, as is the case in the manufacture of such items as fruit juices, tomato ketchup and mayonnaise.

The intensity of sonication is not as high as that of the probe, however, and this leads to some disadvantages:

(i) The ultrasonic power obtained from such a system is limited so that ultrasonically initiated and driven reactions are less likely to be effected by whistle sonication.
(ii) The optimum frequency of blade vibration (designed for 20 kHz) relies upon a rapid pumping of the reaction. Clearly this is not possible for viscous materials.
(iii) The underlying mode of action, involving as it does the flow-induced vibration of a steel blade, immediately suggests that there may be a problem of blade erosion by particulate matter.

4.3.5 Tube reactors

This type of system provides the greatest hope for general usage of ultrasound in the chemical industry. Each of the following systems requires that ultrasonic energy is introduced into a flowing system by vibrations induced in the pipe carrying the reaction media.

(a) The Branson sonochemical reactor [13]

This provides a simple solution to the problem of causing the pipe walls to vibrate. This modular system is made up of units which comprise a pair of 20-kHz probe type sonicators (each 1800 W) mounted on either side of the pipe containing the reaction mixture. The ultrasonic energy from these sources is coupled to the pipe by means of a fluid, which can provide temperature control (Fig. 4.6). Each of these units is therefore quite expensive, in that two complete sonicator systems and the associated plumbing are required. The unit can be used in a batch mode, providing the central pipe section (of 200 cm^3 volume) with an energy of some 15 W cm^{-3}. Alternatively, either singly or in series, the units can be made part of a flow system.

Like all systems based on probes, however, this unit will suffer from erosion of the horn tip and, in the long run, this might prove expensive in terms of 'down-time' for repair and replacement.

A simpler solution to the problem of large-scale flow processing is to ultrasonically vibrate the walls of a pipe through which the reaction flows. In this way the sound energy generated from transducers bonded to the outside of the tube is transferred directly into the flowing liquid. Two design engineering problems are associated with this type of sonicator: (i) the mounting of a transducer on the outer

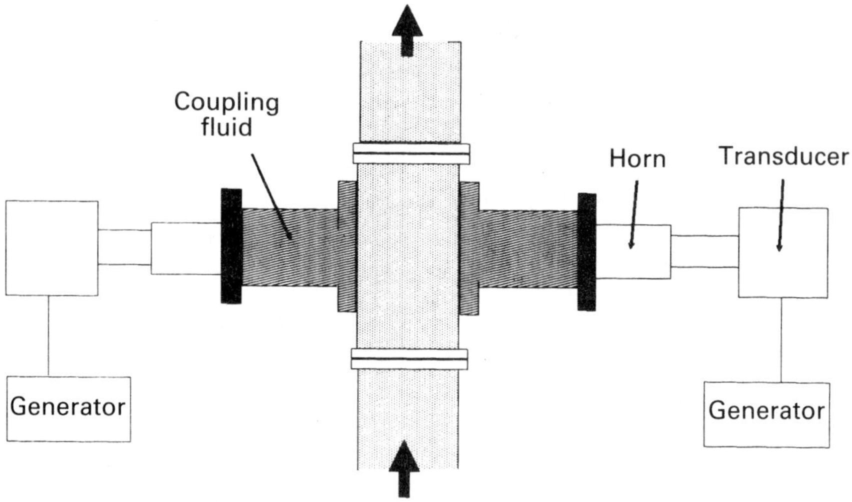

Fig. 4.6 — Branson sonochemical reactor.

tube wall may require precise moulding to a curved surface and (ii) the length of the tube must be such that the ends are at a null point in the sound wavelength in the unit. This will eliminate vibrational problems associated with the coupling of the unit to existing pipework. Generally the commercial tube reactors are constructed of stainless steel; for chemical processing however, other, materials, e.g. titanium, can be used.

(b) Pentagonal cross-section sonicator

In this type of system the engineering problems of bonding the piezoelectric transducers to the pipe surface are removed in that each of the five faces of the pipe is flat.

The dimensions of the Branson Pentagonal Liquid Processor [13] are as shown (Fig. 4.7) giving a volume of 15 l and a quoted throughput of 40 to 190 l min^{-1}. The unit carries 35 piezoelectric transducers, each operating at 35 kHz, and it is claimed that in this configuration the processor provides a uniform ultrasonic field. The smallest dimensions available for such a processor are dictated by the actual size of the flat faces, so that with 40 kHz transducers the cross-section has a minimum value of 3.5 inches (76 mm).

The applications quoted for this type of processor are listed as ultrasonic pateurization, crystallization, degassing, dispersion and emulsification.

(c) Hexagonal cross-section sonicator

A reactor of this type has been developed by Sonic Services of France [14] for the laboratories of J. Berlan in Toulouse (Fig 4.8). In this system the transducers bonded to opposite faces give a 'focus' of energy in the centre of the pipe, unlike the pentagonal style. Such a focus means that very high intensities can be developed

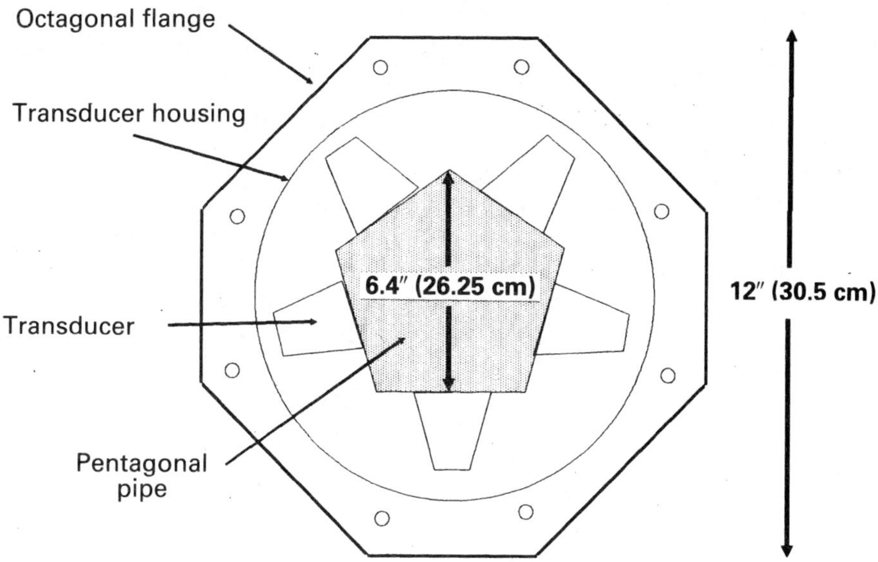

Fig. 4.7 — Pentagonal reactor.

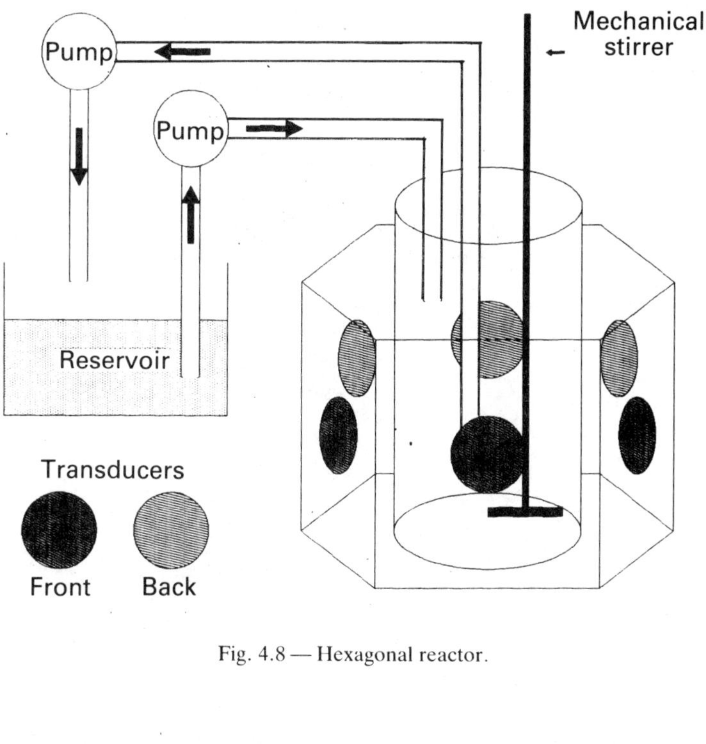

Fig. 4.8 — Hexagonal reactor.

without the associated problems of rapid cavitational erosion on the inner surface of the walls. Additionally there will be no 'shielding' of the vibrational energy entering the system by bubbles developed on the inside walls.

(d) Cylindrical pipe reactor

As with polygon-type tube reactors, the diameter of a cylindrical tube has a minimum value based upon the need to physically bond the transducers to the outer surface; in practice this is an inch or two. Much larger diameter tubes can be used and so high flow rates and viscous materials can be handled with no problems over blockage. In the tubular configuration the ultrasonic energy is focussed toward the middle of the tube with much lower powers at the perimeter inner surface. In this way erosion problems are reduced. In addition to these advantages the system can easily be retro-fitted to existing pipework, thereby requiring minimal disturbance of existing technology.

One of the early types of 'tubular' reactors was developed by Branson for wire cleaning. The transducers are mounted on the outside of a pipe 2.87 inch (73 mm) in diameter, which is sealed inside an outer casing giving the unit an overall dimension of 11 inches (28 cm) long by 10 inches (25 cm) in diameter. This enables the unit to be dipped totally into a cleaning fluid so that any wire to be cleaned can be pulled through the sonicated tube.

A system based on this unit has now become an in-line liquid processor (Fig. 4.9) [13]. Operating at 33 kHz, the system employs 16 transducers, which transfer energy to a cylindrical tube via a coupling fluid. This same coupling fluid can be circulated and refrigerated to provide cooling during the sonochemical processing. Just like the hexagonal reactor, cylindrical flow systems concentrate (focus) the ultrasonic energy in the centre of the flow.

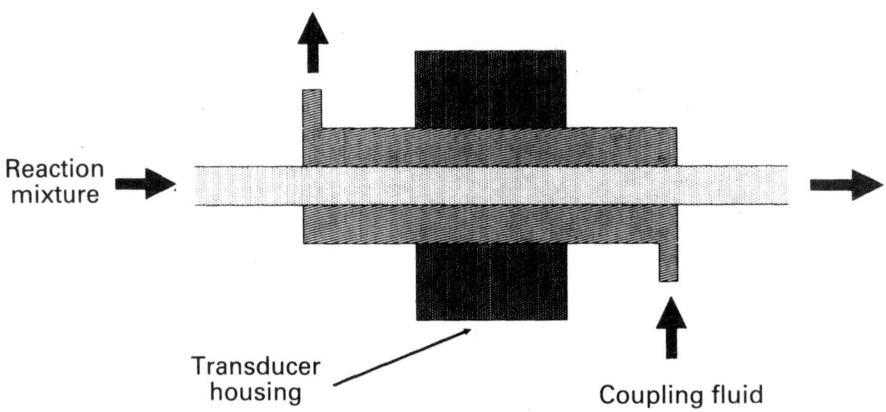

Fig. 4.9 — Cylindrical pipe reactor.

A rather larger resonating pipe 6 inch (152 mm) in diameter operating at 25 kHz has been developed by the Battelle Institute [15] and has found applications in the degassing of liquids.

(e) Cylindrical reactor with core cooling

When very high ultrasonic energy is applied through a pipe, one unavoidable problem is the generation of heat. This has been overcome in the system developed by V. Ragaini in Milan [16]. His cylindrical pipe has a core which consists of a pipe carrying cooling liquid (Fig. 4.10). The annular space between the resonating pipe and the cooling pipe can be used to hold solid reagent or catalyst while the reaction mixture is pumped through the system. In this approach to large-scale sonication the ultrasonic reactor can be seen to be region of uniform but intense sonication through which the reagents are pumped. This is probably a better solution than attempting to sonicate a large vat, which invariably means that a relatively low and non-uniform energy field is involved.

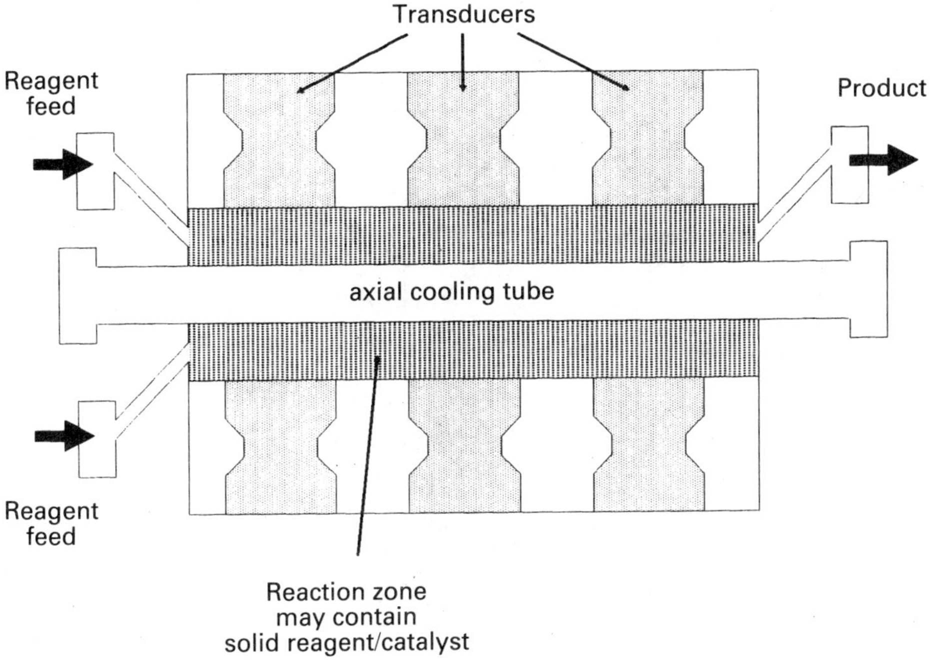

Fig. 4.10 — Cylindrical reactor with core cooling.

The cooling pipe occupies the space which would otherwise be the focus of acoustic energy. In acting as both a temperature control and a co-axial reflector of the sound energy this could well prove to be an efficient design for a flow reactor.

There appears to be no problem in dramatically increasing the ultrasonic energy available for driving such cylindrical tube reactors. Very powerful magnetostrictive transducers suitable for this application are currently being used in engineering to assist in the cold drawing of metal pipes [17]. Sonication of the die at 20 kHz using 6 to 12 kW energy allows for reducing the pipe diameter up to a maximum of 50% per pass, a considerable improvement on conventional methodology. Thus the engineering problems associated with the construction of 'tube reactors' employing such powerful transducers using this principle have already been solved, and so the transfer of this technology to chemical plant should be straightforward.

4.3.6 The RUM system

Low intensity treatment does not automatically mean the use of baths. If the application involves particle size reduction then the choice may well be the 'Reverberatory Ultrasonic Mixing' system (RUM) as developed by the Lewis Corporation in the USA (Fig.. 4.11) [7]. This system can be visualized as two sonicated metal

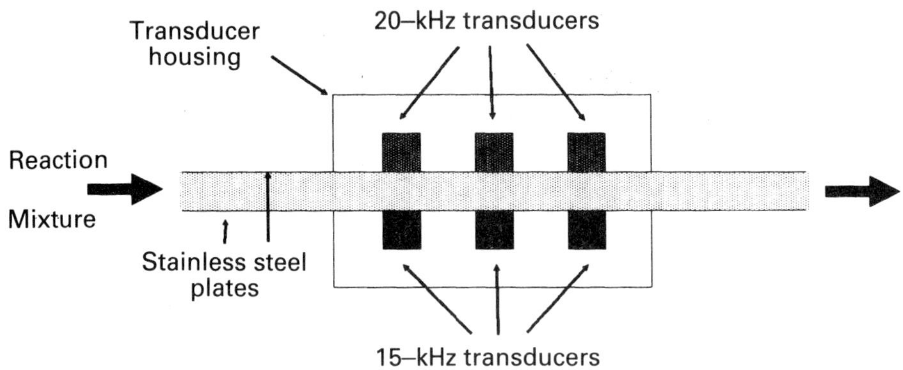

Fig. 4.11 — RUM system.

plates, each similar to the base of a cleaning bath. These enclose a flow system and are driven at different frequencies — normally 15 and 20 kHz. In effect the plates can be regarded as the bases of two ultrasonic baths facing toward each other and separated by only a few centimetres. Under these conditions any liquid flowing between the plates is subject to an ultrasonic intensity greater than that expected from simple doubling of a single plate intensity. The ultrasound 'reverberates' and is magnified in its effect. Already used for applications such as particle size reduction and the extraction of oil from oil shale, it is clearly but a short step to using this system for chemistry. An additional benefit is that, with vibrating plates the size of table tops, the system can cope with very large throughputs of material.

In our laboratories we have shown that the RUM is effective when used for the degradation of polymeric material in solution, e.g. dextran. It also proved particularly useful for the de-agglomeration of solids, e.g. TiO_2 powder.

4.3.7 The Harwell ultrasonic reactor

A few years ago in the UK a research club was set up at the Harwell Laboratories near Oxford called the Sonochemistry Development Club. Funded jointly by a government grant and several private industries, the research has borne fruit in that a flow reactor has been constructed which is capable of generating high intensities with an electrical power rating of 1500 W (Fig 4.12) [18].

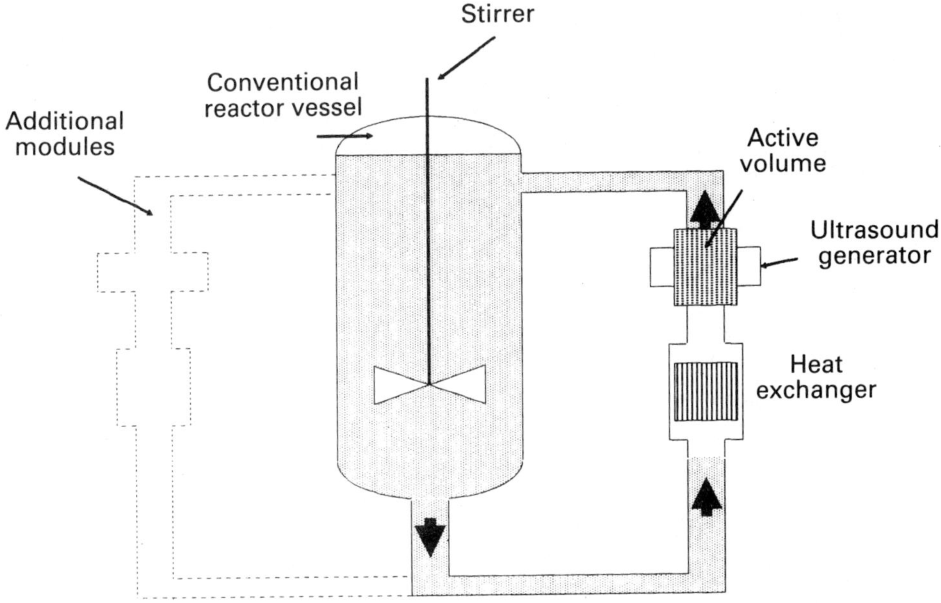

Fig. 4.12 — Harwell sonochemical reactor.

This reactor is said to suffer no cavitational damage to the walls, while providing high energy insonation to the contained liquid. It can be used as a static unit (with a capacity of 4 l) for trials, but is easily incorporated into a flow loop easily handling 50 gallons (230 l) per hour, although the reaction itself may require less residence time and thus be amenable to higher throughputs. Successful chemical reactions performed using this reactor include the Barbier reaction (in which the lithium metal pellets are held in a cage within the sonicated zone) and two-phase liquid hydrolyses. Early results appear to show that with increasing scale-up there is a significant improvement in the efficiency of energy usage.

4.3.8 The Martin Walter push-pull system [19]

A new concept in flow system treatment has been developed in Germany and offers some remarkable possibilities for the chemical industry. The system itself consists of a cylindrical bar of titanium (cut to a precise number of half wavelengths of sound in the material) with opposing piezoelectric transducers bonded to each end and electrically connected through a central hole (Fig. 4.13). With these transducers operating together in a push–pull mode the 'concertina' effect makes the bar of metal expand and contract at half-wavelength distances along the entire length.

The system was originally developed for cleaning the inside of tubes as a submersible transducer, and so it is a sealed unit (Fig 4.14). It can be made to a considerable length simply requiring that the central titanium bar is supported at intervals. One unit which is 10 m long has been developed for the continuous cleaning of a cutting blade of the same dimension.

The benefits of the push–pull system for sonochemistry are obvious. In its simplest form no special plant need be designed since the unit can be slotted co-axially into the centre of existing pipework. More refined systems would involve the design of a flow cell incorporating the push–pull and here some consideration would have to be given to the annular flow dimensions in terms of reflected sound waves from the inner to outer walls. An even more exciting prospect might be to place the push–pull unit inside a tube reactor operating at a slightly different frequency, thereby giving enhanced sonication to the annular flow.

4.4 FUTURE PROSPECTS

At this stage in the development of sonochemistry it seems that the problem is mainly an economic one. If, for each different reaction, a 'tailor-made' reactor was needed to achieve optimum sonochemical yields then equipment manufacturers would need a lot of persuasion to become involved in one-offs. Fortunately, however, although the effects of some variables, including the nature and the shape of the sonoreactor must be very well defined, there is no doubt that such equipment can be derived from the existing selection of systems which are quite wide-ranging in type. For large vat-type reactors, the 'sonic energy' cannot be distributed readily, but this can be partially overcome by the use of smaller sonication cells and the recycling of materials. This is the principle used in the flow reactors.

For ultrasonically assisted reactions the sonochemical yield has very seldom been determined and optimized. For this reason, the economic viability of most of the reported improvements afforded by sonication cannot be established. In a recent paper Suslick presented some encouraging information, by comparing sonochemistry with photochemistry [20] (Table 4.3).

It can be seen that sonochemistry, and more particularly heterogeneous sonochemistry, looks much more attractive than photochemistry as far as the conversion yield of energy is concerned. As photochemical processes are currently used on an industrial scale, sonochemistry should meet with the same success in the near future.

Apart from the technical problems, two other hurdles to sonochemical development are often erected by industrialists: (i) the problem of noise and (ii) the possibility of cavitational erosion in the reactor. It must be acknowledged that all the sonic systems (with the possible exception of liquid whistles) are rather noisy in bench-top operation. But the noise problem is easily overcome by the use of adequate soundproofing, indeed soundproof casings are already used in the case of industrial cleaning baths. On the other hand, the erosion of the reactor is difficult to

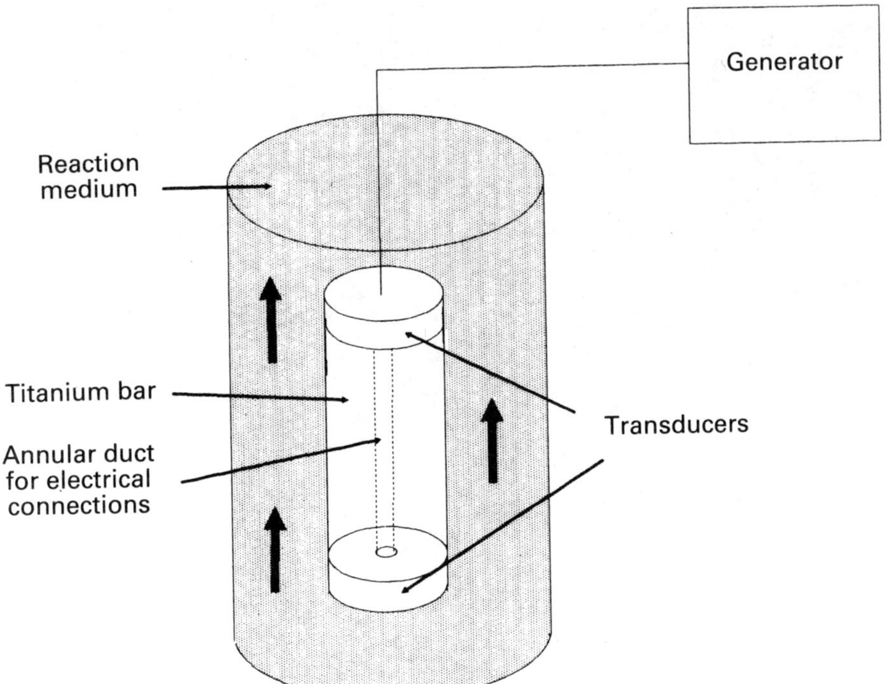

Fig. 4.13 — Martin Walter push–pull system.

avoid. The rate of erosion depends on ultrasonic power, pressure (it increases rapidly with increasing pressure), the nature of the reaction mixture and, in particular, the type of suspended solid. This could be a serious problem with large-scale reactors, but it can be minimized by the use of sets of small continuous flow reactors, which appear to be the preferred research line at present.

Although much remains to be done, the prognosis for the continued health of sonochemistry is good. The combined efforts of chemists, physicists and engineering equipment suppliers should ensure that industrialists have power ultrasound as a viable option in the design of the chemical plants of the future.

REFERENCES

[1] In writing this chapter I would like to acknowledge the help of two researchers from groups who are pioneering the scale-up of sonochemistry: Professor J. Berlan of the Ecole Nationale Superieure d'Ingenieurs de Génie Chimique, Toulouse, France and Dr P. D. Martin of the Sonochemistry Development Club, AEA Harwell Laboratory, UK.

[2] C. J. Martin and A. N. R. Law, *Ultrasonics* (1983) **21**, 85.

[3] B. Pugin, *Ultrasonics* (1987) **25**, 49; see also: T. K. Sakena, *J. Acoust. Soc. of India* (1980) 8; G. Duda and B. Zapior, *Zesz. Nauk.Uniw. Jagiellon., Pr.*

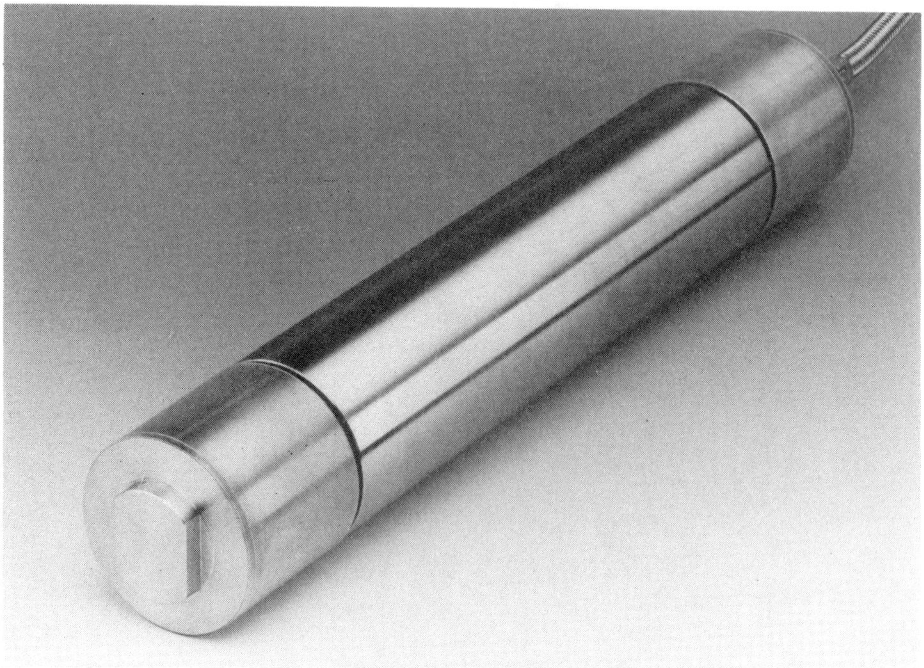

Fig. 4.14 — Photograph of the push–pull processor.

Table 4.3 — Estimations of the relationship between reaction rate and efficiency in terms of power

	Photochemistry	Sonochemistry	
		Homogeneous	Heterogeneous
Rate (mmol min^{-1})	7	1	500
Yield (mmol kWh^{-1})	2	5	200

Chem., (1983) **28**, 43. The propagation of acoustic waves is a very complex problem: P. M. Morse, *Vibration and sound*, McGraw-Hill, 1948; P. M. Morse and K. U. Ingard, *Theoretical acoustics*, McGraw-Hill (1968).

[4] F. G. P. Aerstin, K. D. Timmerhaus and H. S. Fogler, *AIChE Journal* (1967), **13**, 453.

[5] J. Berlan, Sonochemistry Meeting of Leatherhead Food Research Association and of the Sonochemistry Group of the Royal Society of Chemistry, Leatherhead, 10 May (1990).

[6] B. Brown and J. E. Goodman, *High intensity ultrasonics*, Iliffe Books, 1965.
[7] Lewis Corporation, 102 Willenbrock Road, Oxford, Connecticut 06483, USA.
[8] Dawe Ultrasonics, 27 Clayton Road, Hayes, Middlesex UB3 1AN, UK.
[9] R. W. Wood and A. L. Loomis, *Phil. Mag.* (1927) **4**, 417.
[10] W. Janovski and R. Pohlmann, *Z. Angew. Phys.* (1948) **1**, 22.
[11] R. E. Singiser and H. M. Beal, *J. Am. Pharm. Assoc., (Scient.Ed.)* (1960) **49**, 482.
[12] R. S. Davidson, A. Safdar, J. D. Spencer and D. W. Lewis, *Ultrasonics* (1987) **25**, 35.
[13] Branson Ultrasonics, Eagle Road, Danbury, Connecticut 06810-1961, USA.
[14] Sonic Services, 23 rue Pascal, 75005 Paris, France.
[15] Battelle, Columbus Laboratories, 505 King Avenue, Columbus, Ohio 43201, USA.
[16] Professor V. Ragaini, Universita di Milano, Dipartimento di Chimica Fisica ed Elettrochimica, Via Golgi, 19-20133 Milano, Italy.
[17] Technoform Sonics, Enterprise Trading Estate, Pedmore Road, Brierley Hill, West Midlands DY5 1TX, UK.
[18] Harwell Sonochemistry Development Club, AEA Industrial Technology, Harwell Laboratories, Oxford OX11 0RA.
[19] Martin Walter Ultraschalltechnik GmbH, Hardstrasse 13 - Postfach 6, Ortsteil Conweiler, D-7541 Straubenhardt 5, Germany.
[20] K. S. Suslick, *A.C.S. Symposium Series,* **333**, 123.

5

Sonochemistry experiments

In this chapter a number of representative sonochemistry experiments have been gathered together. These have been supplied from the laboratories credited with each experiment and I am most grateful to those colleagues who have helped in compiling the list. The broad divisions within this chapter have been into types of ultrasonic equipment used and subdivisions then deal with reaction types.

5.1 EXPERIMENTS USING ULTRASONIC BATHS

ORGANOZINC REACTIONS

5.1.1 The synthesis of β-cuparenone

from: J. L. Luche, C. Petrier, A. E. Greene *and* J. P. Lansard, *L.E.D.S.S., Chimie Recherche, Universite Scientifique et Medical de Grenoble, France.*

Introduction

The conjugate additions of alkyl, alkenyl and aryl moieties to unsaturated substrates constitutes a very important synthetic procedure [1]. However there are limitations, especially in the case of hindered substrates where the very general methods using copper derivatives are frequently useless. Other reagents have been proposed, among which organozinc derivatives are of special value [2]. Generally these reagents are not conveniently prepared from the metal and an indirect synthesis has to be effected from an orgonolithium or magnesium compound and a zinc salt. It is because of the simplicity of preparation of organolithium reagents under sonication that a one-pot preparation of diorganozincs can be envisaged, and a typical example of this approach is described below [3] (Scheme 5.1).

Method

A round-bottomed Pyrex flask (25 cm^3) was charged with 4-bromotoluene (347 mg, 2.03 mmol), anhydrous zinc bromide (225 mg, 1.00 mmol) and lithium wire (28 mg, 4.00 mmol) in 6 cm^3 of dry ether. After bubbling argon through for 3–5 min, the flask was placed in the optimum position (as described in section 2.3.1) of a Sonoclean

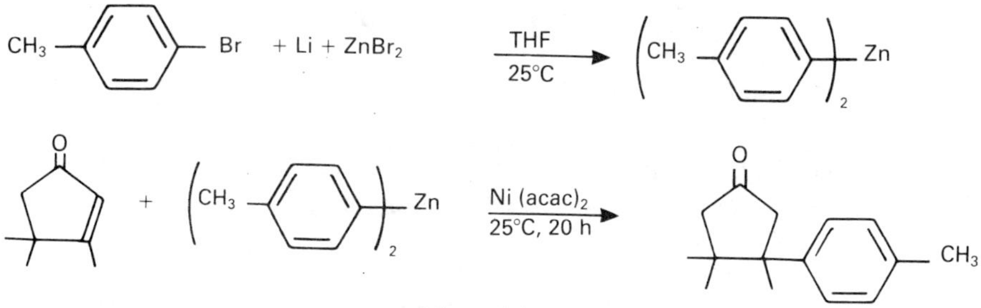

Scheme 5.1

SHE 2500 ultrasonic cleaner. Sonication was effected at room temperature for 15 min. Attack of the lithium proceeded very rapidly and an intense black colour developed throughout the solution.

A mixture of 3,4,4-trimethylcyclopent-2-en-1-one (62 mg, 0.5 mmol) and nickel acetylacetonate (3 mg, 0.01 mmol) in 2 cm^3 of ether was then added, and the mixture was magnetically stirred at room temperature for 20 h. Hydrolysis was effected by pouring the black suspension into saturated aqueous NH_4Cl, and the organic products were extracted with ether. The ether extract was washed (aq. NaCl), dried over Na_2SO_4 and the solvent evaporated. The residue was chromatographed on silica gel to give 72 mg (67%) of pure β-cuparenone.

Discussion

The present method of synthesis of β-cuparenone involves the formation of a quaternary carbon atom vicinal to another quaternary centre. All the attempts to effect the reported conjugate additions with organocopper derivatives failed [4], or gave the undesired 1,2-addition. Diorganozinc reagents are able to give much better yields of the desired 1,4-addition, but they are not readily accessible. The sonochemical preparation constitutes a highly valuable alternative which can be used in a variety of cases. The conventional process under stirring can also be successful, but in general the yields are 20% lower and the reaction time exceeds 1–2 h. In the case described here, Wurtz–Ullmann coupling to the biaryl is virtually absent, which makes the purification of the compound very easy.

The preparation of diaryl zinc reagents can be effected without difficulty in a cleaning bath, but for reasons which are not at all clear this experimental system was not capable of generating dialkyl compounds. A much more powerful source of ultrasound (probe system) was required for these preparations and this is described later (Experiment 5.2.9).

References

[1] G. H. Posner *An introduction to synthesis using organocopper reagents*, Wiley Interscience, New York, 1980.

[2] C. Petrier, J. C. de Souza Barboza, C. Dupuy and J. L. Luche, *J. Org. Chem.* (1985) **50**, 5761.

[3] A. E.Greene, J. P. Lansard, J. L. Luche and C. Petrier, *J. Org. Chem.* (1984) **49**, 931.
[4] A. Casares and L. A. Maldonado, *Synth. Commun.* **(1976) 6**, 11.

5.1.2 Sonochemical conjugate additions in aqueous media: the preparation of a key intermediate in the synthesis of side-chain hydroxylated metabolites of vitamin D3

from: J. L. Mascarenas, L. Castedo and A. Mourino, *Departamento de Quimica Organica y Seccion de Alcaloides del CSIC, Santiago de Compostela, Spain.*

Introduction

Conjugate addition to α,β-unsaturated carbonyl compounds is most frequently effected using the popular organocopper reagents. The use of other organometallic methods requires the protection of sensitive functionalities and is incompatible with some functional groups (e.g. the free hydroxyl of an alcohol) [1]. Carrying out conjugate additions in an aqueous medium, by sonochemical techniques is of interest for several reasons. Firstly, it is much easier, from a practical point of view, to effect a reaction in an aqueous medium than in one which has been carefully dried. Secondly, as the presence of hydroxyl groups in the solvent has no inhibitory effect, the substrate can accommodate free hydroxyl groups. Thirdly, the process proves to be safe, rapid, and highly efficient. A typical example which demonstrates all of these advantages is described is shown in Scheme 5.2.

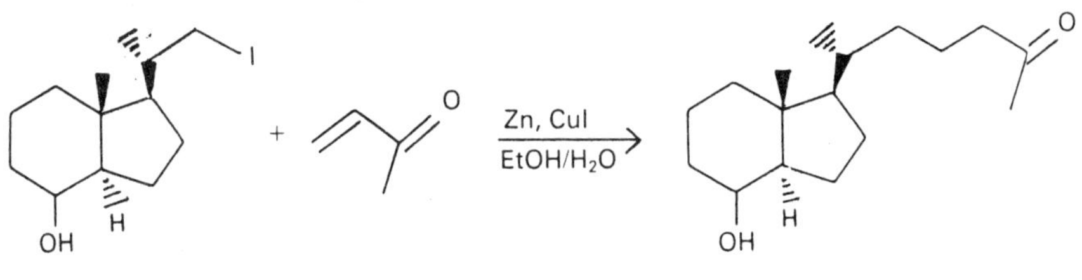

Scheme 5.2

Method

Sonication was afforded by a standard Bandelin KH102 cleaning bath (frequency 35 kHz and nominal power 150 W) and the location of the position of maximum energy was accomplished by the normal method. The temperature was kept constant by addition of crushed ice (the removal of water effectively kept the level in the bath constant). Round-bottomed Pyrex reaction tubes (length 12 cm, diameter 2.5 cm), with gas inlet side arm were used without further special adaptation.

To the sonicated solution of de-A,B-23,24-dinor-22-iodocholan-8β-ol (140 mg, 0.44 mmol) [2] and freshly distilled methyl vinyl ketone (65 μl, 55 mg, 0.8 mmol) in deoxygenated EtOH:H_2O (7:3, 6 cm^3), were added Zn dust (110 mg, 1.7 mmol) [3] and CuI (84 mg, 0.44 mmol) [4]. The initial grey mixture turned rapidly black and was sonicated under an argon atmosphere for 20 min, before adding more Zn (55 mg, 0.85 mmol) and CuI (42 mg, 0.23 mmol). The reaction was monitored by thin-layer chromatography (TLC) [5]. After further sonication for 30 min, the reaction was judged complete and Et_2O (10 cm^3) was added. The resulting mixture was filtered through celite and the solids washed with Et_2O (3×30 cm^3). The filtrate was washed with brine (20 cm^3) and the aqueous phase extracted with additional Et_2O (15 cm^3). The combined organic layers were dried over $MgSO_4$, filtered and concentrated under reduced pressure on a rotary evaporator. The crude product was purified by flash chromotography [6] (5% ethyl acetate/hexane) to afford, after concentration at reduced pressure and high vacuum drying, 87 mg of de-A,B-8β-hydroxy-27-norcholestan-25-one (76%) [7] as a colourless oil.

Discussion

A few years ago, Luche and coworkers [8a,b,c] developed an efficient method for the addition of alkyl halides to electron-deficient alkenes in aqueous media, using sonochemical methods. The reaction described here is an application of this method. The product of the synthesis above is a key intermediate in the synthesis of a side chain of 25-hydroxyvitamin D3 and 1α,25-dihydroxyvitamin D3, the more important metabolites of vitamin D3.

The advantages of this synthesis is the short reaction time and the removal of both the need to use dry solvents and the need to employ low temperatures, as required in previous syntheses [9a,b,c]. This synthetic method does not effect other functional groups in the molecule and, in this particular case, avoids the need to protect the hydroxyl group at C-8. The classical method of preparation also uses four steps, with an overall yield of only 60%. The normal method of preparation of the zinc–copper couple takes several hours whereas the sonochemical method gives an almost immediate result.

The procedure described here could be extended to a wide variety of haloalkanes and electron-deficient alkenes for the rapid, inexpensive, and easy synthesis of different natural products in high yields.

References

[1] G. H.Posner, *An introduction to synthesis using organocopper reagents*, Wiley: New York, 1980.

[2] (a) L. Castedo, J. L. Mascarenas, A. Mourino and L. A. Sarandeses, *Tetrahedron Lett* (1988) **29**, 1203; (b) J. L. Mascarenas, L. A. Sarandeses, L. Casedo and A. Mourino, *Tetrahedron* (1991) **47**, 3485.

[3] The zinc dust was purified by successive washing under magnetic stirring with 5% HCl (3 min), H_2O (10 min), saturated NH_4Cl (15 min), H_2O (5 min), and EtOH (3 h). The solid was filtered, washed with EtOH and ethyl acetate, and dried in a desiccator over P_2O_5: L. F. Fieser and M. Fieser, *Reagents for organic synthesis*; John Wiley 1967; vol 1, p. 1292.

[4] The CuI was purified by dissolution in a boiling saturated aqueous solution of KI, filtration and precipitation by addition of cold water. The white solid was

filtered, washed with H_2O, ethyl acetate, and Et_2O, and dried in a desiccator over P_2O_5: G. B. Kauffman, L. A. Teter, *Inorg. Synth.* (1963) **7**, 9.

[5] TLC conditions: 20% ethyl acetate/hexane; 1: R_f 0.55; 2: R_f 0.40; de-A,B-23,24-dinorcholan-8β-ol (from the hydrogenation of 1): R_f 0.60.

[6] Flash chromotography was performed using silica gel (Merck 60, 230–400 mesh), according to the method of Still: W. C. Still, A. Mitra, *J. Am. Chem. Soc.* (1987) **100**, 1927.

[7] J. L. Mascarenas, A. Mourino and L. Castedo, *J. Org. Chem.* (1986) **51**, 1269. ^{1}H NMR (250 MHz, $CDCl_3$, p.p.m.) 0.91 (3H, d, J=6.76 Hz, CH_3-21), 0.92 (3H, s, CH_3-18), 2.13 (3H, s, CH_3CO), 2.38 (2H, m, CH_2CO), 4.07 (1H, m, 0.5 W=3 Hz, CH-8); IR (neat) 3610, 1720, 1120 cm^{-1}; MS (EI, 70 eV) *m/e* 266(M+, 0.2), 251(0.5), 248(0.5), 163(1.5), 135(8), 111(56), 95(18), 81(34), 79(29), 67(50), 55(100).

[8] (a) C. Petrier, C. Dupuy and J. L. Luche, *Tetrahedron Lett.* (1986) **27**, 3149; (b) J. L. Luche and C. Allavena, *Tetrahedron Lett.* (1988) **29**, 5369; (c) J. L. Luche, C. Allavena, C. Petrier and C. Dupuy, *Tetrahedron Lett.* (1988) **29**, 5373.

[9] (a) B. Lythgoe, D. A. Roberts and I. Waterhouse, *J. Chem. Soc. Perkin Trans. I* (1977) 2608; (b) G. A. Leyes and W. H. Okamura, *J. Am. Chem. Soc.* (1982) **104**, 6099; (c) J. L. Mascarenas, A. Mourino and L. Casedo, *J. Org. Chem.* (1986) **51**, 1269.

5.1.3 Large-scale cyclopropanation reaction

from: O. Repic, *Sandoz Research Institute, E. Hanover, NJ 07936, USA.*

Introduction

In 1982 Repic described a sonochemical modification of the Simmons–Smith reaction using sonochemically activated zinc which avoided the sudden exotherm normally associated with this type of reaction [1]. Up to this time most existing methods for this reaction relied upon activation of the zinc by using zinc–silver or zinc–copper couples and/or the use of iodine or lithium. In the sonochemical procedure no special activation of the zinc was required, indeed equally good — and reproducible — yields were obtained using zinc in the form of dust, foil, mossy or even rods. The ultrasonic source employed was a small laboratory cleaning bath.

This methodology was successfully scaled up to achieve the cyclopropanation of methyl oleate in 0.5-kg quantities as described below (Scheme 5.3).

Scheme 5.3

Method

Apparatus.

The reactor (shown in Fig. 5.1) consists of a 22-l four-necked resin flask equipped with three Claisen adapters to fit two spiral condensers, a mechanical stirrer, thermometer, nitrogen gas inlet, gas outlet, an additional funnel and two stainless steel rods (equipped with stainless steel bearings, e.g. SGA Scientific Inc. catalog JS-1160), from each of which a cone of zinc was hung with copper wire.

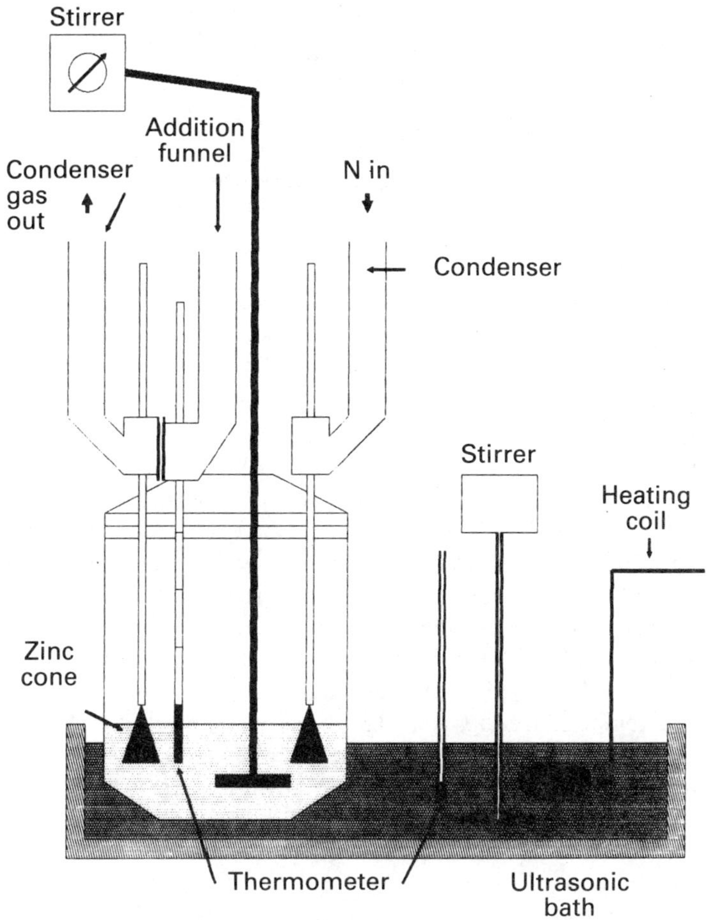

Fig. 5.1 — Large-scale cyclopropanation apparatus.

A cork stopper was mounted on each steel rod just above the bearing; the cork hole was made tight enough to hold the rod and the zinc cones in any vertical position. A clamp holder was attached on top of the steel rod as a safety stop. Each cone was manufactured by melting c. 900 g of zinc into a 125-cm^3 conical flask, positioned over an acetylene gas burner. When the metal cooled, the flask was tapped onto the bench, the bottom of the flask fell off, and the zinc was recovered in one cone-shaped piece. Four holes were drilled through the neck of the cone, and copper wire was laced through the holes. The reactor was immersed (with some difficulty, owing to its buoyancy) into a 36×18×18 inch (46×91×91 cm), 50-gallon (230-l) ultrasonic bath (3000 W, 80 kHz, model Z-5000 S, manufactured by Zenith Manufacturing and Chemical Company). The bath was filled with water–ethylene glycol (1:1), stirred with an air-driven propeller–stirrer and heated with two 1600-W electrical immersible heating coils, as well as by a 4000-W electrical heater built into the bath. The experiment was conducted inside the walk-in hood. The gaseous by-products of the reaction were vented from the top of the condenser into the hood exhaust.

Chemicals
The zinc was obtained in the form of bars from Alfa (8×8×300 cm, 99.9%); 1,2-dimethoxyethane (99%) and Red-Al were purchased from Aldrich; diiodomethane (99%) was distilled prior to use.

Procedure
The reactor was charged under nitrogen with 1,2-dimethoxyethane (2700 cm^3) and two zinc cones (total mass 1800 g, 27.5 mol) were immersed into the solvent to about 3 cm below the top of the cone. The stirrers, the heaters, the ultrasonic irradiation and a timer (time=zero) were turned on and left on throughout the experiment. After some 30 min (reaction temperature 30°C), a stock solution of Red-Al (72 cm^3, 0.017 mol; prepared by diluting 42 cm^3 of commercial 3.4-M solution in toluene with 1,2-dimethoxyethane to 600 cm^3) was added rapidly. At 60 min from time zero (reaction temperature 55°C) a solution of iodine (5 g, 0.019 mmol) in 30 cm^3 of 1,2-dimethoxymethane was added rapidly. At 130 min (reaction temperature 85°), the zinc cones were raised above the solvent surface, and diiodomethane (1288 cm^3, 16 mol) was added rapidly, taking care to avoid splashing the zinc. At 150 min (reaction temperature 86°C), methyl oleate (593 g, 2 mol) was added rapidly. At 155 min (reaction temperature 89°C), the zinc cones were lowered into the solution. A vigorous reaction and reflux set in, with minimal foaming. (CAUTION: exothermic reaction and evolution of flammable gases!) The mixture was irradiated with ultrasound and heated at 100±2°C for 2.25 h (until 290 min from time zero) at which time gas–liquid chromatography (GLC) indicated complete reaction (less than 1% of methyl oleate remaining) [3].

At this point the ultrasound and the heaters were turned off and the zinc cones were raised above the solution surface. They were later weighed and it was found that 806.5 g (12.3 mol) of zinc had reacted. Large amounts of ice were added to the bath to cool it to −15°C. The reaction mixture was diluted with toluene (2 l) and the solid zinc iodide was dissolved by slowly adding cold 2-M hydrochloric acid (4 l) (CAUTION: exothermic!). The clear solution was transferred to a 22-l separatory funnel. The aqueous layer was separated (8 l) and extracted with two 2-l portions of toluene. The organic layers were combined (8 l) and washed with two 2-l portions of water, one 0.5-l portion of saturated aqueous sodium chloride, two 1-l portions of saturated aqueous sodium thiosulphate, two 1-l portions of water until pH=6, and one 0.5-l portion of saturated aqueous sodium chloride. The organic layer was evaporated at 50°/60 mm to obtain a yellow oil containing some toluene, 2% of methyl *trans*-9,10-methyleneoctadecanoate and less than 1% of methyl oleate. This crude product was distilled at 150–155°C/0.3 mm (85% yield) and filtered through silica gel (adsorbent–sample weight ratio 5:1) with dichloromethane as the solvent to remove most of the impurities. Thus 510 g of methyl *cis*-9,10-methyleneoctadecanoate (97% pure, 82% yield) was obtained as a light yellow oil [4].

Discussion

The method has several advantages over the normal method of cyclopropanations as a result of changing from zinc powder to the metal:

(i) There is a reduction in foaming (normally associated with ethene and cyclopropane formation),
(ii) the exotherm is more evenly distributed (only a small clean area of metal is available throughout the reaction),
(iii) the reaction can be controlled by removing the lump of metal from the reaction (this is not unlike the use of fuel rods in a nuclear reactor),
(iv) the residual metal can be removed from the reaction as a lump.

References

[1] O. Repic and S. Vogt, *Tetrahedron Lett.* (1982) **23**, 2729.
[2] O. Repic, P. G. Lee and U. Gigers, *Org. Preps and Procedures Int.* (1984) **16**, 25.
[3] The GLC conditions were 2 m×2 mm 3% OV-1 on 80/100 Supelcoport, helium 30 cm^3 min^{-1}; temperature programme: 200°C→300°C at 10°/min; retention times: methyl oleate, 5.6 min; methyl 9,10-methyleneoctadecanoate, *trans*: 6.0 min; *cis*: 6.5 min.
[4] IR (neat): 1740 (ester), 3050, 1010 (cyclopropane) cm^{-1}; 1H-NMR ($CDCl_3$): 3.65 (s, 3H, methoxy), 2.3 (t, 2H, methylene), 1.0–1.7 (m, 26H, methylenes), 0.9 (t, 3H, methyl), 0.6 (m, 3H, *cis*-cyclopropane), −0.3 (m, 1H, cyclopropane H *cis* to R) (Me_4Si=0.0); ^{13}C-NMR ($CDCl_3$): 174.20 s, 51.34 q, 34.07 t, 31.90 t, 30.17 t, 30.10 t, 29.66 t (2C), 29.42 t, 29.32 t, 29.27 t, 29.15 t, 28.69 t, 28.64 t, 24.94 t, 22.66 t, 15.75 q, 15.70 d, 14.07 d, 10.89 t. ($CDCl_3$=77.00)

ORGANOLITHIUM REACTIONS

5.1.4 The Wurtz coupling of bromobenzene using lithium metal

from: The Coventry Polytechnic Sonochemistry Group.

Introduction

This is an ideal experiment for the demonstration of the effect of ultrasound on organometallic chemistry. It is interesting for the student because various colour changes occur during this reaction, making it visually appealing. The work was originally reported by Boudjouk [1] but, perhaps owing to the use of a more effective bath for our experiments, the reaction times have been reduced considerably [2] (Scheme 5.4).

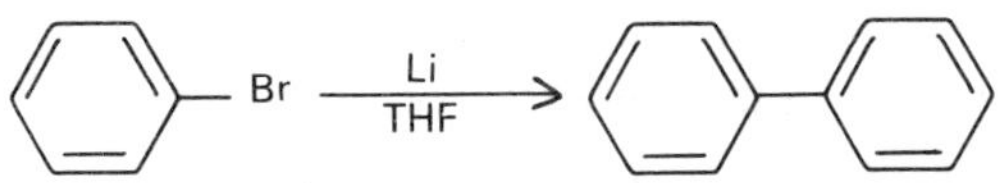

Scheme 5.4

Method

Bromobenzene (15.7 g, 0.1 mol), lithium shot (0.75 g, 0.11 mol, washed with petroleum ether to remove oil), naphthalene (0.3 g, added to assist organolithium formation) and tetrahydrofuran (25 cm^3) were placed in a 50-cm^3 quickfit conical flask equipped with a condenser (as a precaution against any excessive exotherm).

The flask was placed in an ultrasonic bath (Kerry Pulsatron 60, 40 kHz, 50 W) filled to within 1 cm of the top with water containing 2% Decon 90. The flask was immersed to a depth of a centimetre or two such that the reaction could be readily observed and maximum agitation of the contents was achieved. The level of sonication of the flask contents takes several forms depending on position. Ideally the surface should be sufficiently disturbed to 'shoot' small jets of solution upwards and against the walls of the vessel. Sometimes this is accompanied by visible 'pulsing' of vapours in the upper flask. At other stages the reaction mixture may be seen to seethe or rotate violently. If little or no surface activity is visible the flask should be repositioned and, in fact, this must be done several times during the reaction as the ultrasonic bath water warms up.

In the first few moments a blue colour is seen to stream from the surface of the lithium metal which starts to colour the whole solution. Within a minute the reaction becomes blue/green and after about 2 min it becomes quite hot (although it does not achieve reflux temperature) and the reaction mixture turns dark brown. Thereafter the reaction temperature falls and the reaction itself turns dark green after about 12 min. Sonication must be maintained through the total reaction time of about 30 to 40 minutes and the end of the reaction is signalled by a change in reaction colour from dark green to caramel brown.

At the end of the reaction the mixture is filtered through glass wool to remove any residual lithium and then quenched by the addition of a few drops of water. After the tetrahydrofuran had been removed from the orange liquid by rotary evaporation, crystallization of the residue afforded crude biphenyl which was purified by recrystallization from methanol. The yield is typically around 4 g (55%).

References

[1] B. H. Han and P. Boudjouk, *Tetrahedron Lett.* (1981) **22**, 2757.

[2] T. J. Mason, J. P. Lorimer and J. Moorhouse *Education in Chemistry* (1989) **26**, 13.

5.1.5 Preparation of lithium diisopropylamide (LDA)

from: J. Einhorn *and* J. L. Luche, *L.E.D.S.S., Chimie Recherche, Université Scientifique et Medical de Grenoble, France.*

Introduction

It is unnecessary to underline the importance of alkyl lithium reagents in organic synthesis, amongst which primary, secondary, and tertiary butyl lithiums are probably the most widely used [1]. They are nucleophiles and bases, and can deprotonate a variety of compounds with an acidic hydrogen via a hydrogen-metal exchange. The formation of a molecule of butane or isobutane renders these processes irreversible. Ultrasonic irradiation facilitates their preparation [2], but their manipulation remains somewhat difficult. Because of their extreme sensitivity to air and moisture, and their reactivity towards many compounds, they have to be prepared in hydrocarbon solvents and their solutions manipulated with specialized apparatus, especially when using large quantities. In order to ameliorate these difficulties and take advantage of sonochemical methods of synthesis, we considered the possibility of the *in situ* preparation of organolithiums [3].

Strong bases, such as alkyl lithium reagents, exhibit a marked nucleophilic character; thus the deprotonation of ketones, esters, acids, etc., is frequently impaired by the undesired formation of adducts. The lithium salts of secondary amines have been found to be non-nucleophilic strong bases, and one of the most widely used is LDA [4]. This highly useful compound is prepared rapidly and with great ease using sonochemical methods (Scheme 5.5).

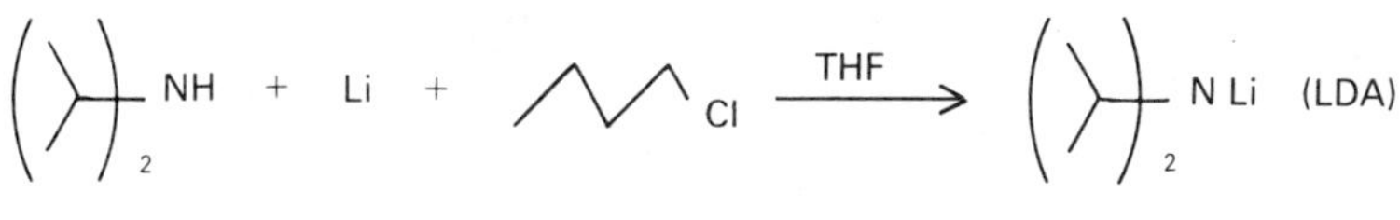

Scheme 5.5

Method

The experiments were carried out in a simple cleaning bath (Sonoclean SHE 2500, 150 W, 45 kHz, or Kerry Ultrasonics, Pulstron 250, 220 W, 38 kHz) filled with water and cooled to 15–18°C by circulation of tap water in a stainless steel coil.

The reaction flask (round bottom, single neck, Pyrex, 50 cm^3) was charged with lithium wire (0.140 g, 20 mmol, Aldrich), 1-chlorobutane (0.926 g, 10 mmol) and diisopropylamine (0.010 g, 10 mmol, freshly distilled from potassium hydroxide) in dry tetrahydrofuran (THF) (10 cm^3). The flask was stoppered with a rubber septum, flushed with argon and sonication started. The metal very quickly became silvery and then progressively disappeared during which time the liquid turned pale yellow. After consumption of the metal (15–30 min) the strength of the LDA solution was standardized by the normal method [5]. Reproducible yields between 90 and 95% of the theory can be obtained.

The same experiment was scaled up to 0.5 mol as follows. A 500-cm^3 round-bottom, single-neck, Pyrex flask was charged with 1-chlorobutane (46.3 g, 0.5 mol), diisopropylamine (50.6 g, 0.5 mol) and dry THF (250 cm^3). After clamping in the ultrasonic bath, lithium (7 g, 1 mol) was added by portions of *c*. 1 g over a 90-min period, under an argon blanket. Progressive addition is necessary to avoid excessive heating due to the exothermic reaction. After dissolution of the metal (120 min overall), the yellowish solution containing precipitated lithium chloride was standardized to give a yield of 91%.

Discussion

This preparation of LDA is a typical example of the use of sonochemical techniques in synthetic chemistry. It is accomplished in a short time without technical difficulty by simply mixing all the reagents. In contrast to the usual procedure, no solvent exchange is necessary and the operation does not necessitate low temperatures. However, it also illustrates one of the present limitations of sonochemistry. Cleaning baths used in this work deliver a limited energy. As the rate-limiting step in this process is most probably the ultrasonically enhanced electron transfer from the metal to the carbon-chlorine bond [6], too low an energy results in an increased reaction time when the sample size is increased. We have found that experiments run on quantities larger than 0.5 mol require much longer irradiation time, and undesired side-reactions appear.

The procedure described here can be extended to the preparation of various non-nucleophilic strong bases such as lithium bis(trimethylsilyl)amide and lithium 2,2,6,6-tetramethyl piperidide with quasi-quantitative yields.

Two further one-pot syntheses involving deprotonation by organolithium reagents are given below.

References

[1] B. J. Wakefield *The chemistry of organolithium compounds*, Pergamon Press, Oxford. 1974.

[2] For a review, see: J. Einhorn, C. Einhorn, J. L. Luche, *Synthesis* (1989) 787.

[3] J. Einhorn and J. L. Luche, *J. Org. Chem.* (1987) **52**, 4123.

[4] M. Fieser and J. C. Smith, *Reagents for Organic Synthesis* Wiley, Vol 13, 1988 p.163.

[5] In 1 cm^3 of dry THF, 0.5 mmol of pure, carefully weighed menthol is dissolved with a small crystal (less than 0.5 mg) of orthophenanthroline. The mixture, cooled to −78°C, is treated with a solution of LDA, slowly added until a persistent red colour appears.

[6] J. C. de Souza Barboza, J. L. Luche and C. Petrier, *Tetrahedron Lett.* (1987) **28**, 2013.

5.1.6 Preparation of the dianion of isobutyric acid — a reaction 'cascade'

from: J. Einhorn *and* J. L. Luche, *L.E.D.S.S., Chimie Recherche, Université Scientifique et Medical de Grenoble, France.*

Introduction

The solutions of bases obtained as described above (section 5.1.5) are used for the deprotonation of various substrates. Because of a limited stability, they have to be employed as rapidly as possible. An improvement of the method, in order to mimimize the possible degradations, would be to generate the lithium reagents in the presence of the acidic compound. The preparation of the highly useful dianion of a carboxylic acid [1] provides an example of this approach (Scheme 5.6).

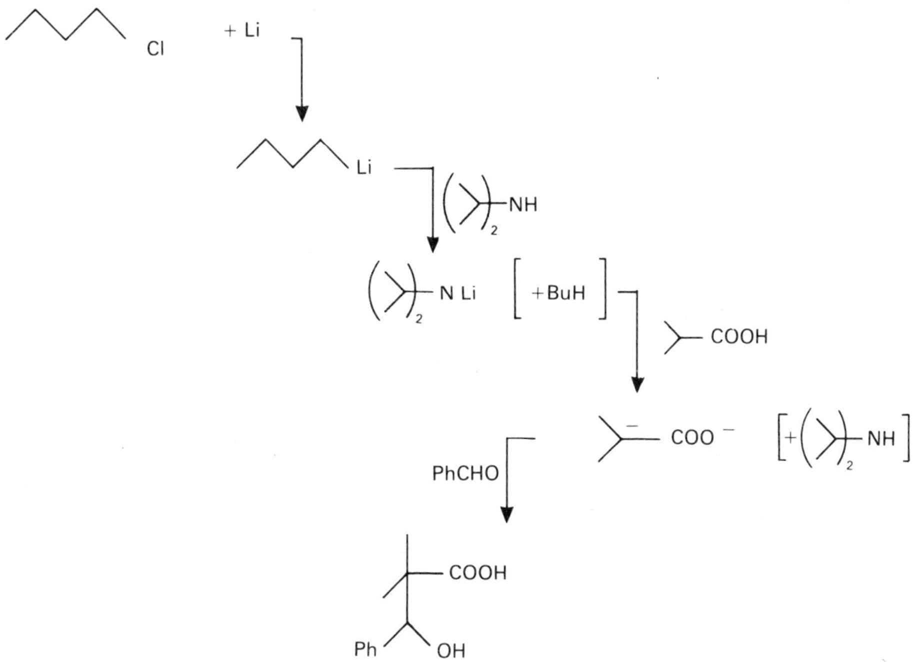

Scheme 5.6

Method

Using the sonication set-up described above, a Pyrex flask (25 cm^3, single neck) was charged with lithium wire (0.140 g, 20 mmol), 1-chlorobutane (0.926 g, 10 mmol), diisopropylamine (1.010 g, 10 mmol) and 2-methylpropanoic acid (0.440 g, 5 mmol) in dry THF (10 cm^3). As soon as sonication was started, a rapid reaction occurred during which a thick white precipitate appeared — the lithium salt of the acid. This salt progressively redissolves to give a light yellow solution of the dianion after 20 min. Benzaldehyde (0.53 g, 5 mmol) was added via a syringe and the mixture stirred at 20°C for 1 h. After hydrolysis in water (20 cm^3), the mixture was extracted with ether (2×10 cm^3) and then the aqueous phase was acidified with 2-M HCl and extracted again. The ether phase was dried over Na_2SO_4 and the solvent evaporated to yield a solid residue which was recrystallized from hexane–ethyl acetate to yield 3-hydroxy-2,2-dimethyl-3-phenylpropanoic acid 0.757 g (78%), m.p. 132–133°C (literature value 133–134°C) [8].

Discussion

These reactions constitute a cascade, in which the role of ultrasonic waves is essential. In the absence of ultrasound the initial complex mixture consisting of an amine, an alkyl halide, an acid and lithium metal follows a complex series of reactions in which several compounds are found. To avoid these problems, and to ensure that the sequence follows our picture of a cascade, it is essential that the first stage — 'upper reservoir' — is filled as rapidly as possible. In other words, it is necessary to form the alkyl lithium reagent in a fast step. This can be achieved with the help of sonic waves.

References

[1] For a review, see N. Petrignani and M. Yonashiro, *Synthesis* (1982), 521.
[2] C. R. Hauser and D. S. Breslow, *J. Am. Chem. Soc.* (1939) **61**, 793.

5.1.7 Preparation of 2-(1,3-dithianyl)lithium

from: J. Einhorn *and* J. L. Luche, *L.E.D.S.S., Chimie Recherche, Université Scientifique et Médical de Grenoble, France.*

Introduction

The anion of 1,3-dithiane has been used extensively in organic synthesis, as the precursor of various compounds, essentially aldehydes and ketones, because of its easy alkylation by alkyl halides and tosylates [1]. It is generally prepared by deprotonation of the parent heterocycle with butyl lithium. In the continuation of the ideas developed in the preceding experiments, it was obviously of interest to effect this preparation by sonochemical means (Scheme 5.7).

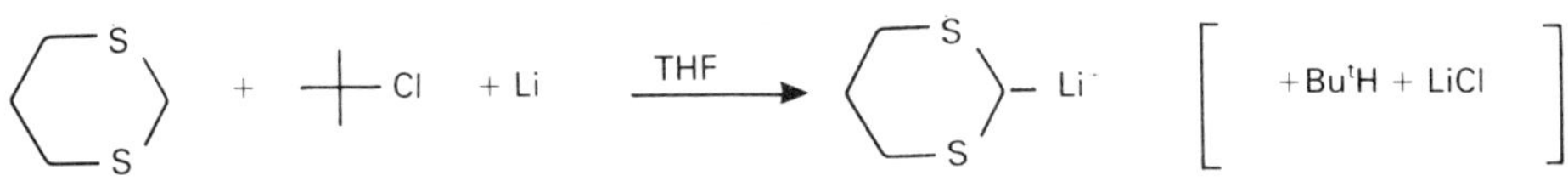

Scheme 5.7

Method

Using the experimental set-up as previously described, a reaction flask (10 cm^3, round bottom, Pyrex), filled with 2-chloro-2-methylbutane (93 mg, 1 mmol), 1,3-dithiane (120 mg, 1 mmol) and lithium sand (14 mg, 2 mmol) in dry THF (2 cm^3) was sonicated until the metal disappeared (20 min). The colourless solution was treated with benzaldehyde (106 mg, 1 mmol), stirred for 1 h, then hydrolysed and worked up as by the normal method. The solid residue was recrystallized from hexane to give 225 mg of the dithiane adduct (98% yield). m.p. 73–74°C (literature value 71–72°C) [1].

Discussion

This reaction constitutes an example of the *in situ* preparation of *t*-butyl lithium. This reagent is extremely sensitive to air and water (spontaneously inflammable). The *in situ* generation offers then the valuable advantage that we can avoid the manipulation and accumulation of this reagent, since it is consumed as it is formed.

It is necessary, however, to underline two important points. This synthesis definitely requires the use of *t*-butyl chloride (2-chloro-2-methylpropane). Although *n*-butyl lithium is certainly able to deprotonate the intitial 1,3-dithiane, the resulting anion is reactive towards 1-chlorobutane itself. Substantial amounts of 2-butyl-1,3-dithiane are thus formed; this undesirable pathway is avoided when steric hindrance slows down the nucleophilic displacement process. The second point is that it is essential to use lithium containing 2% sodium, as no reaction occurs from pure lithium (containing less than 0.02% Na) even under ultrasonic activation. The origin of this effect is not known [3]. It is interesting to note that physical phenomena such as surface passivation can be in part responsible for this situation, but are certainly not the only ones involved. Complex chemical reactions, such as electron transfers, seem to be of major importance to account for the role played by minor amounts of sodium [3].

References

[1] For a review, see E. J. Corey and D. Seebach, *Angew. Chem. Int. Ed. Engl.* (1965) **4**, 1075.

[2] B. J. Wakefield, *The Chemistry of Organolithium Compounds*, Pergamon Press, Oxford, 1974, p. 24.

[3] For a hypothesis, see J. L. Luche, C. Petrier, C. Einhorn, J. Einhorn, J. C. de Souza Barboza, C. Dupuy, P. Delair, C. Allavena and T. Tuschl, *Ultrasonics* (1990) **28**, 316.

ORGANOTIN REACTIONS

5.1.8 Addition of allylic groups to carbonyl compounds — preparation of 4-methyl-4(2-propenyl)butanolide

from: C. Einhorn *and* J. L. Luche, *L.E.D.S.S., Chimie Recherche, Université Scientifique et Médical de Grenoble, France.*

Introduction

The addition of various allylic groups to carbonyl compounds has been widely studied [1]. This reaction is generally performed with allylic organometallics of lithium, magnesium, zinc, etc. New procedures using derivatives of boron, silicon, titanium and tin, have also been proposed for better stereoselectivities, especially for open chain aldehydes and ketones [2]. We reported that in the case of these latter substrates, the reaction can be efficiently achieved in a quite unusual aqueous medium consisting of water and an organic co-solvent such as THF. In the presence of zinc powder and aqueous saturated ammomium chloride solution, a rapid, spontaneous reaction occurs with allyl chloride or bromide and an aldehyde or a ketone, to give the homoallylic alcohol in good yield [3]. It was found later that changing zinc for tin leads to a better yield and an improved tolerance of the system towards various functional groups. In this situation, however, the metal is much less reactive than zinc and the mixture has to be activated by sonication [4]. An illustration of the synthetic possibilities is given here (Scheme 5.8).

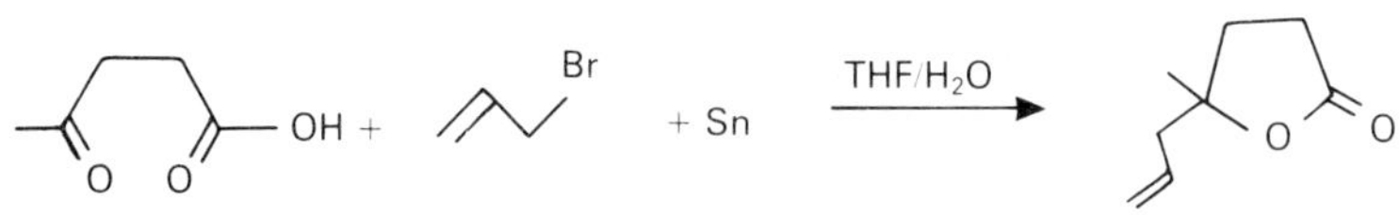

Scheme 5.8

Method

Sonication was performed in a cleaning bath using the methodology described above (section 5.1.7). The allylic halide and the metal are always used in equimolar amounts. A mixture of levulinic acid (224 mg, 2 mmol), allyl bromide (484 mg, 4 mmol) and tin powder (472 mg, 4 mmol) were added to 1 cm^3 of THF and 1 cm^3 of water. Sonication was effected at 15–18°C until the grey colour of the metal had disappeared and a thick white precipitate had been formed (*c*. 45 min). The mixture was diluted with THF and filtered through celite. The solid was washed with THF and the organic phases were washed with saturated aqueous NH_4Cl, dried (Na_2SO_4) and the solvent evaporated to give an oil which was purified by column chromatography on silica gel. The yield was 200 mg (71%) of an oily compound [5,6].

Discussion

When compared with many methods considered as classical, the present allylation process exhibits several advantages: it is accomplished in a short time, the reagents are neither exotic nor difficult to prepare and it is compatible with functionalities which usually have to be protected (e.g. COOH), or would otherwise interfere. Sonication is necessary in all cases involving tin metal, as the reaction mixture does not react under simple stirring, and the parallel method using zinc gives a lower yield in a number of cases [4].

In this reaction the mechanism of the action of ultrasound is probably twofold. The classical mechanical role probably contributes to the increased reactivity of the solid reagent, especially with a soft metal such as tin. Ando has suggested that the lattice energy of the solid, more or less represented by its melting point, could be an important factor [7]. Soft materials like tin are more prone than others to undergo the erosion and particle fragmentation effects of ultrasound. The other mechanistic possibility, the direct chemical effect, stimulating the electron transfer from the metal to the organic halide, should also be considered. In this case, however, it is not easy to demonstrate it unambiguously.

References

[1] R. W. Hoffman, *Angew. Chem. Int. Ed. Engl.* (1982) **21**, 555.
[2] Y. Yamamoto, H. Yatagai, Y. Ishibara, N. Maeda and K. Maruyama, *Tetrahedron* (1984) **40**, 2239.
[3] C. Petrier and J. L. Luche, *J. Org. Chem.* (1985) **50**, 910.
[4] C. Einhorn and J. L. Luche, *J. Organomet. Chem.* (1987) **322**, 177.
[5] R. Couffignal and A. Tougani, *C. R. Acad. Sci. Paris* (1984) **298**, 73.
[6] IR (neat): 3060, 1770, 1640, 1200, 1160, 1000, 940 cm^{-1}; ^1H NMR (80 MHz, $CDCl_3$, TMS internal reference): 6.2–5.5 (m, 1H), 5.3–4.9 (m, 2H), 2.8–1.8 (m, 6H), 1.4 (s, 3H) p.p.m..
[7] T. Ando and T. Kimura, Ultrasonic organic synthesis involving non-metal solids, in *Advances in sonochemistry*, Vol 2, T. J. Mason Ed, J.A.I. Press, 1991, in press.

COLLOIDAL POTASSIUM

5.1.9 Carbon–sulfur bond cleavage of sulfolane

from: T.-S.Chou, *Institute of Chemistry, Academia Sinica, Taipei, Taiwan.*

Introduction

Sonication using a cleaning bath is a simple and convenient method for preparing colloidal alkali metals [1]. For example the blue colour typical of colloidal potassium is produced in a few minutes when the metal is insonated in toluene or xylene. These colloidal systems are particularly useful in condensations of the Dieckmann type, in the generation of ylide intermediates in Wittig reactions and in the desulfurization of sulfones [2]. It is this last use of the colloidal metal which is shown in the following reaction (Scheme 5.9).

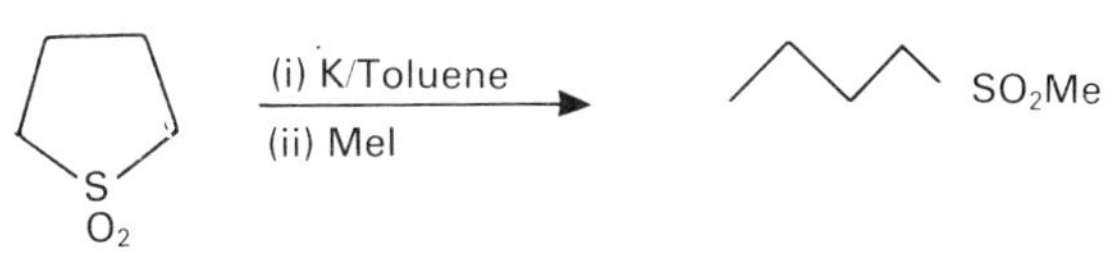

Scheme 5.9

Method

A cylindrical Pyrex reaction vessel was used which was 8 cm in height and 3.5 cm in diameter. The vessel was flat-bottomed for effective ultrasonic transmission. To the oven-dried vessel containing anhydrous toluene (10 cm^3, distilled from potassium) was added finely cut potassium metal from which the surface oxides had been carefully removed (292 mg, 7.5 mmol). The metal fragments were pressed with a glass rod into pellets and the reaction vessel was subjected to ultrasonic irradiation in a cleaning bath [3]. Detergent was added to the water in the cleaning bath and the level of the water was about 4 cm from the bath top. Sonication was continued for about 10 to 15 min until the main portion of metal was dispersed to a blue colloid.

The position and the depth of the reaction vessel in the cleaning bath was adjusted to optimize the agitation in the reaction. This position was near the central point of the bath but the depth needed to be finely adjusted for different runs. The bath temperature was kept roughly between 25 and 40°C by adding a small amount of crushed ice when necessary.

A sulfolane (300 mg, 2.5 mmol) in toluene (10 cm^3) was added dropwise to the reaction, during which time the potassium became completely dispersed, and the reaction mixture was irradiated for a further 10 min. The reaction was quenched by the dropwise addition of moisturized THF (3.8 cm^3, prepared by dissolving 1.8 g of H_2O in 50 cm^3 of THF) over a period of 20 min. During this period the blue colour gradually became light blue, light green, and finally yellow. The reaction mixture was transferred to a round-bottomed flask containing CH_3I (1.6 g, 12.5 mmol) and THF (20 cm^3) and the resulting mixture was allowed to reflux for 2 h. The excess of THF was removed under reduced pressure and the crude oil extracted with CH_2Cl_2 (4×10 cm^3). The combined organic layers were washed with 10% $Na_2S_2O_3$, dried ($MgSO_4$) and concentrated under reduced pressure to give essentially pure butyl methyl sulfone (91%).

Conclusion

In the absence of ultrasound the reaction becomes very sluggish and more side products are obtained. The reductive cleavage of the C–S bond of dialkyl sulfones is difficult to achieve by other reducing agents such as Raney nickel, sodium amalgam and lithium in amines.

References

[1] J.-L. Luche, C. Petrier and C. Dupuy, *Tetrahedron Lett.* (1984) **25**, 753.
[2] T. S. Chou and M. L. You, *Tetrahedron Lett.* (1985) **26**, 4495.
[3] Three types of commercially available ultrasonic cleaning baths have been used satisfactorily in our laboratory: Transonic T570/H (frequency 35 kHz from Elma, Germany); Sonorex RK 100H (frequency 35 kHz from Bandelin Electronics, Germany); Bransonic 220 (frequency 35 kHz from Branson, USA).

5.1.10 Removal of SO_2 from *cis*-2,5-diheptyl-3-sulfolene

from L. T.-S. Chou, *Institute of Chemistry, Academia Sinica, Taipei, Taiwan.*

Introduction

The stereospecific extrusion of SO_2 from a sulfolene (Scheme 5.10) can be achieved by thermal reaction (heated at about 130°C) or by treatment with lithium aluminium hydride for 2 h. The advantage of the ultrasonic method described below is that it is very mild and efficient [1].

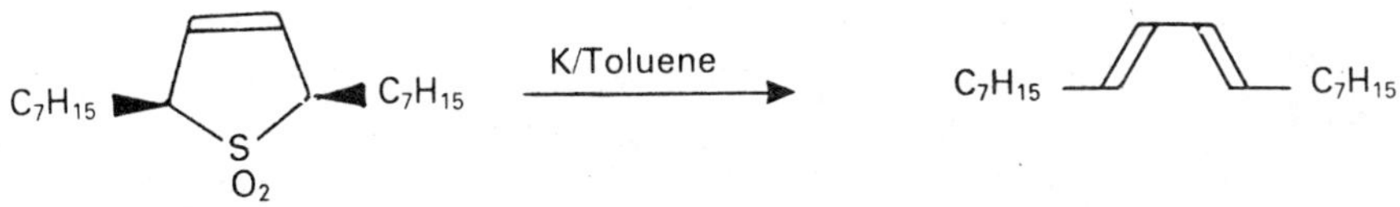

Scheme 5.10

Method
cis-2,5-Diheptyl-3-sulfolene (314 mg, 1 mmol) in anhydrous toluene (3 cm^3) was added dropwise to a suspension of ultrasonically dispersed potassium (97 mg, 2.5 mmol) in toluene (4 cm^3) at room temperature. The ultrasonically dispersed potassium is prepared by the same procedure using the same apparatus as in section 5.1.9. After the mixture was irradiated with ultrasound for 1 min, the reaction mixture was allowed to stand until the colloidal potassium settled, the clear solution was drawn out using a syringe and concentrated under vacuum to give spectroscopically pure dienyl product in 92% yield.

Conclusion
The stereospecificity of this reaction suggests that it takes place via a concerted process of cheleotropic extrusion of sulfur dioxide.

Reference
[1] T. S. Chou and M. L. You, *J. Org. Chem.* (1987) **52**, 2224.

SONOELECTROCHEMISTRY

The effects of ultrasound on electrochemical processes remain a largely unexplored area, but research suggests significant benefits. These include modifications to the chemistry of reactions at the electrode and greatly increased current efficiencies. One major result of these studies could be that electrochemistry on an industrial scale may become more viable. The experiments below show some of these possible advantages.

5.1.11 Convenient and selective sonoelectrochemical synthesis of the anions Se_2^{2-}, $Se^{2-}Te_2^{2-}$ and Te^{2-}

from: C. Degrand, *Laboratoire d'Electrochimie Organique, Université Blaise Pascal, 63177 Aubiere cedex, France.*

Introduction
Ultrasound, applied via a simple cleaning bath, has been shown to be beneficial in the electrochemical synthesis of Se_2^{2-} and Se^{2-} ions. The method shows considerable advantages over previous techniques used for the preparation of these anions involving electro-reduction with a sacrificial Se electrode. A further major advantage of the sonoelectrochemical method is that it is also applicable to the synthesis of the corresponding tellurium anions (Scheme 5.11) [1].

$$Se_2 \xrightarrow{e} Se_2^{2-} \xrightarrow{e} Se^{2-}$$

$$Te_2 \xrightarrow{e} Te_2^{2-} \longrightarrow Te^{2-}$$

Scheme 5.11

Method

The ultrasonically promoted electrochemical reduction of insoluble Se or Te powder was performed in an H-type cell, the three compartments of which are separated by glass frits or by ion-exchange membranes Ionac MA 3475 (anodic side) and MC 3470 (cathodic side) and filled with an aprotic solvent (DMF, CH_3CN, THF) containing 0.1-M Bu_4NBF_4 or Bu_4NPF_6 (0.2-M in the case of THF) (Fig. 5.2; for simplicity the glass tubing for argon and N_2 bubbling, and the compartment covers are not shown). The volume of the cathodic, central and anodic compartments are 60, 25 and 60 cm^3, respectively. The cathode is a graphite cloth of cylindrical shape (Carbone Lorraine, Ref TGM 389) or a Pt grid, and the anode is a Pt grid. The reference electrode is an aqueous saturated calomel electrode (SCE) immersed in a separate compartment (glass tubing with fritted glass) (Tacussel, Ref PDL 1) filled with the solvent saturated with Bu_4NBF_4 or Bu_4NPF_6).

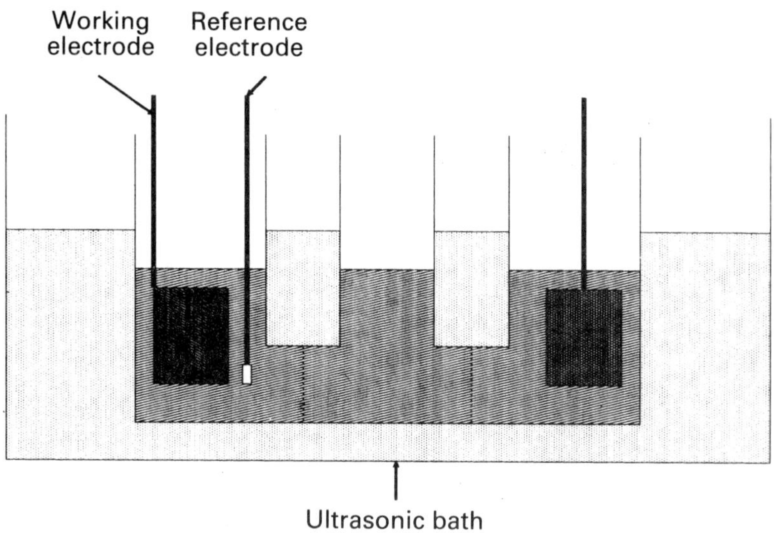

Fig. 5.2 — Cell for electrochemical synthesis.

The cathodic solution was de-aerated with argon prior to the introduction of Se or Te powder, and an inert atmosphere of argon was then carefully maintained. The central and anodic compartments of the cell were de-aerated with N_2 prior to and during electrolysis. The cell was partially immersed in an ultrasonic bath (Bransonic B32), which was cooled with ice during the electrolysis. The electrolysis was carried out at controlled potentials with an Amel 552 potentiostat and a Tacussel IG5-N integrator (coulometry). The coulometric monitoring of the electrolysis indicates the nature of the electrogenerated species. As long as less than one electron per Se or Te unit is consumed, the Se_2^{2-} and Te_2^{2-} dianions are generated, whereas the cathodic synthesis of Se^{2-} and Te^{2-} involves two electrons per Se or Te unit.

Reference

[1] B. Gautheron, G. Tainturier, and C. Degrand, *J. Amer. Chem. Soc.* (1985) **107**, 5579.

5.1.12 Electrochemical synthesis of a diphenyl diselenide

from: C. Degrand, *Laboratoire d'Electrochimie Organique, Université Blaise Pascal, 63177 Aubiere cedex, France*

Introduction

The Se and Te anions produced electrochemically in section 5.1.11 can be used *in situ* for the synthesis of selenides and tellurides in good yield (60 to 90%) via nucleophilic displacement from haloalkanes (Scheme 5.12) [1].

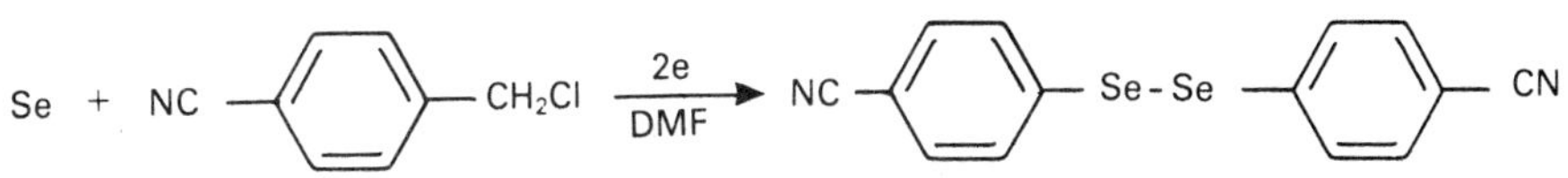

Scheme 5.12

Method

The cell described above (Fig. 5.2) contained analytical-grade DMF carefully dried on neutral alumina and 0.1-M Bu_4NPF_6. Se powder (395 mg, 5 mmol) was introduced and its reduction was carried out under sonication at −0.65 V. The cathodic solution turned brown (owing to the formation of Se_2^{2-} anions). To maintain a reasonable Faradaic current (about 120 mA), the working potential had to be progressively moved to more negative values (from −0.65 to −0.9 V). The electrolysis was stopped after the consumption of 193 C (2 mmol electrons) then 4-cyanobenzyl chloride (2 mmol) was added and sonication maintained for a further 15 min.

The catholyte was diluted with water (600 cm^3) and the electrolysis products extracted with diethyl ether (3×200 cm^3). After the ethereal solution had been dried with $MgSO_4$, the solvent was removed and the crude product analysed by ^1H NMR.

It contained the selenide (0.90 mmol; 90%) and traces of 4-cyanobenzyl chloride (*c*. 2%). The diseleno derivative (m.p.=140°C) was recrystallized from methylene chloride/hexane.

Conclusion

Sacrificial Se and Te electrodes have been used, prior to this technique [2]. However, they involve the fusion of a mixture of Se and graphite, or the use of a Te stick, at high temperatures.

A large number of synthetic methods have been developed to prepare seleno-anionic species. A convenient and widely used technique is the reduction of Se by trialkylborohydrides, but these are relatively expensive materials [3]. In tellurium chemistry, the number of synthetic reactions is rather small [4], and commercially available Te is known to exhibit a low chemical reactivity, owing to its surface oxidation.

References

[1] G. Tainturier, B. Gautheron and C. Degrand, Organometallics (1986) **5**, 942.
[2] J. A. Gladysz, J. L. Hornby and J. E. Garbe, *J. Org. Chem.* (1978) **43**, 1204.
[3] C. W. Kirk, D. Nalewajek, G. B. Blanchet, H. Schaffer, F. Moraes, R. M. Boysel and F. Wudl, *J. Am. Chem. Soc.* (1985) **107**, 675.
[4] G. Merkel, H. Berge and P. Jeroschewski, *J. Prakt. Chem.* (1984) **326**, 467.

5.1.13 Kolbe electrolysis of cyclohexanecarboxylate

from: The Coventry Polytechnic Sonochemistry Group.

Introduction

A classical and long-established electrochemical synthesis is the formation of alkanes (R–R) from the electrodecarboxylation of carboxylate salts (e.g. $RCOO^-Na^+$), the so-called Kolbe reaction. The mechanism governing the formation of the dimeric product (R-R) is a single anodic electron transfer (SET) process leading to carboxylate radicals, which, after decomposition to alkyl radicals by loss of carbon dioxide, dimerize (Scheme 5.13).

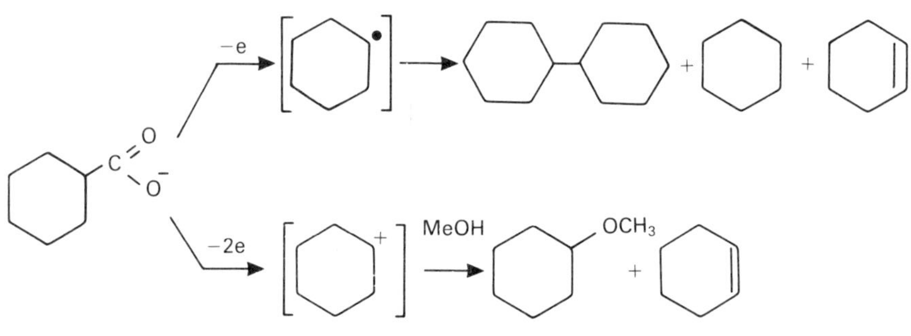

Scheme 5.13

Alternative fates for these radical species are either disproportionation to yield equal proportions of alkene and alkane or abstraction of a hydrogen atom from a solvent molecule to yield an alkane. The key to the success of this dimerization reaction is the formation of the radicals close to the anode surface. A major competing reaction to SET occurs when the alkyl radical above undergoes a second oxidative electron transfer to give a carbocation (i.e. an overall two-electron-per-molecule process). Subsequent reactions of this carbocation lead to the formation of both substitution and elimination products. The ratio of the products from one- and two-electron processes depends on a variety of factors, which include the structure of carboxylate and such experimental parameters as current density, reaction solvent and electrode material.

Method

In a typical procedure [1] cyclohexane carboxylic acid (2 g, 0.0156 mol) was dissolved in 40 cm^3 of methanol (B.D.H., Analar). To this was added 3.55 cm^3 of tetramethyl ammonium hydroxide (25% in methanol) and the resulting solution was made up to 50 cm^3 with methanol. Prior to the start of the electrolysis the solution was degassed by purging the solution with N_2 for 20 min, and during the electrolysis N_2 was blown just above the surface. This solution was electrolysed in the presence of ultrasound. The electrolysis was carried out using two platinum flag electrodes (active area of each electrode 1 cm^2) with a separation distance of approximately 3–5 mm. A potential was applied across the electrodes using a Thurlby PL310 power source (both the voltage applied and electrolysis current were displayed on the power supply) to produce an electrolysis current of 200 mA cm^{-2}. The average potential required for this current density is 8.3 V in the absence and 7.3 V in the presence of ultrasound.

The electrolysis current was passed through the solution for 5 h, ensuring completion of the oxidation reaction. Ultrasound was provided by immersing the vessel containing the solution into a Kerry Pulsatron 55 Ultrasonic bath (40 kHz, 50 W). The reaction vessel was cooled by adding small amounts of crushed ice to the bath water.

On completion the reaction was extracted with diethyl ether and the products analysed by GLC using a Perkin-Elmer F17 instrument with a 10% Carbowax 20-M column (see Table 5.1). The identity of each product was verified by comparison with authentic materials.

Discussion

It is an empirical observation that under normal electrolysis conditions the one-electron route is favoured by the use of methanol as solvent, platinum electrodes and by acidic (i.e. partially neutralized) systems. Our results for the electro-oxidation of cyclohexanecarboxylate in the absence of ultrasound are in general accord with this observation in that, although products from both one-electron and two-electron pathways were formed, the product from the former process, bicyclohexyl, predominated (Table 5.1). The minor products, cyclohexene (4.8%) and methoxycyclohexane (2.8%), may be explained in terms of the two-electron process, where cyclohexene is generated via proton loss and methoxycyclohexane from solvent (methanol)

Table 5.1 — Electrolysis of cyclohexanecarboxylate[a,b]

	No ultrasound	Ultrasound
Bicyclohexyl	49.0	7.7
Ccyclohexane	1.5	2.6
Cyclohexene	4.5	32.4
Methoxycyclohexane	2.1	34.3
Methyl cyclohexanoate	17.0	2.5

[a]Relative product ratios by GLC after the passage of 2.2 F $mole^{-1}$.
[b]Average cell potential to maintain current density of 200 mA cm^{-2} is 8.3 V in the absence of ultrasound and 7.3 V in its presence.

attack on the intermediate cyclohexyl carbocation. Methyl cyclohexanoate is probably generated in the reaction through simple esterification.

In the presence of ultrasound, the product yields were changed quite significantly. For example bicyclohexyl, the major product in SET, fell to 7.7%, while the ether and cyclohexene both dramatically increased from their yields under conventional conditions — 2.1% to 34.3% and 4.5% to 32.4% respectively. This suggests that the two-electron process is favoured under ultrasonic irradiation to the extent of *c.* 65% and the SET process contributes less than 10% to the total reaction under these conditions. Further, a lower overall cell voltage was required to maintain current density during the sonochemical reaction (7.3 V versus 8.3 V) and the reaction attained completion in a shorter timespan, despite the greater preponderance of the two-electron process. This corresponds to a greater current efficiency under ultra-sonic irradiation, which could be associated with greater ion mobility produced in the region of the electrode possibly coupled to a suppression in solvent discharge. It seems likely that both of these factors could provide a contribution to the ultrasonically induced change in mechanism.

It should be noted that, under the influence of ultrasonic irradiation, the platinum electrode remained bright and reflective throughout the reaction, with no fouling evident.

Reference

[1] A. Chyla, J. P. Lorimer, T. J. Mason, G. Smith and D. J. Walton, *J. Chem. Soc. Chem Commun.* (1989) 603.

5.1.14 Sonochemically enhanced electroless plating of nickel

from: The Coventry Polytechnic Sonochemistry Group.

Introduction

The benefits of electroplating under the influence of an ultrasonic field have been established over the last 30 years or so, and it was in the 1950s that several papers were produced which showed the real potential of sonically assisted electroplating

[1]. The simplest way of introducing ultrasonic irradiation during the course of plating is by sonication of the plating bath itself. Such irradiation was found to increase the deposition rate and to negate the fall in plating current which normally occurs owing to polarization [2]. In the case of nickel plating over a wide current density range it was claimed [3] that the presence of ultrasonic irradiation alleviated the need for rigorous control of bath temperature in order to obtain acceptable deposits. Improvements in the efficiency of chromium plating was also reported at that time [4]. Careful study revealed an improvement in the quality of the plating itself in terms of both adhesion and hardness, although the use of very high current densities was accompanied by an increase in the porosity of the plate.

Our own interest in this field is in electroless nickel plating, which is a process involving the chemical deposition of nickel in the absence of an electric current. Electroless nickel is a low temperature coating process which, after heat treatment, produces plating with a hardness comparable to values achieved by other surface hardening processes, such as nitriding. One major benefit of this type of plating is that electroless nickel has the ability to plate uniformly even on complex shapes; this is certainly not the case with electrolytic coatings, which tend to build up on corners or other irregularities.

Rich has investigated the effects of ultrasonic irradiation (20 kHz) on the electroless deposition of nickel and discovered that the coating rate of an alkaline hypophosphite electroless plating bath was increased fivefold [5]. More recently, Mallory has shown that similar rate enhancements can be achieved with acid electroless nickel baths and that the hardness of the deposits was increased by the use of ultrasound [6]. Additionally, the phosphorus content of these deposits was observed to be lower than that of coatings produced without ultrasound.

Method

The metal samples required for plating were made from plain low-carbon steel sheet. Each sample was of approximate dimensions 25×25×3 mm. Pretreatment of these samples was important to ensure successful application of the electroless nickel coating since inadequate cleaning results in a lack of adhesion, roughness, coating porosity and early failure of the coat. The steps involved in pretreatment were as follows:

(a) Degrease in 1,1,1,-trichloroethane and air dry.
(b) Soak for 10 min in 50% aqueous Decon 90 (detergent) at 60–80°C. Rinse in distilled water.
(c) Electroclean at between 3 and 5 Am^{-2} for 90 s. The current was supplied by a Coutant LA 200.2 D.C. supply. Three reversals of current were employed by a manual switch box, with the final reversal being anodic. The stainless steel electrodes employed were passivated in 30% HNO_3 for 24 h. Rinse in distilled water.
(d) Dip in 30% HCl for 30 s to activate the nucleation sites. Rinse in distilled water.
(e) Electroclean again at 3 to 5 $A\,m^{-2}$ for 45 s. Finally rinse in distilled water.

After the final rinse of the pretreatment the specimen was transferred to a 250 cm^3 glass beaker (soaked in 30% HNO_3 for 24 h prior to plating) containing

200 cm^3 of an acid hypophosphite reduced plating solution stabilized with urea. The samples were suspended in the solution on nichrome wire.
Three methods of plating were employed;

(a) plating in a still solution,
(b) plating in a magnetically stirred solution and
(c) plating in a sonicated bath.

In the sonochemical method the glass beaker containing the plating solution was suspended in a heated Kerry (model KS 351 HT) cleaning bath (frequency 38 kHz). The ultrasound was applied 1 min after the sample was dipped in the plating bath. The temperature of the solution was monitored using a thermocouple. At the end of the plating process, the samples were removed from the solution, washed with distilled water and air dried. To condition the final coating the samples were heat-treated at 400°C for 1 h. To avoid oxidation of the samples, a Wild Barfield VMR 2 vacuum furnace was employed.

The thickness of the deposit was measured using a micro 2000 electronic micrometer. Measurement was taken at the centre of each sample to an accuracy of ∓1 μm. The microhardness of deposits was established using a Buehler Micromet digital microhardness tester. Samples were subjected to test loads of 50 and 100 g for a load time of 15 s.

At 90°C the rate of conventional and stirred electroless nickel plating was 10 and 20 $\mu m\ h^{-1}$ respectively. Ultrasonic irradiation dramatically increased the rate to 50 $\mu m\ h^{-1}$. The heat treated ultrasonically plated sample had a microhardness of 1055 VPN (Vickers Pyramid Number) whereas that of the stirred plating was 1022 VPN.

Discussion

In addition to rate enhancement, it would seem possible that economically desirable changes to the bath composition can be achieved by the use of ultrasonic irradiation. For example while Minjer and Brenner [7] observed that addition of thiourea to an acid electroless nickel bath at very low concentrations increased the plating rate towards a maximum (with further addition completely inhibiting the plating reaction), Mallory observed that, with ultrasound, the concentration of thiourea necessary to cause these changes was considerably less than otherwise required. Under simultaneous irradiation the requirement for this stabilizer is reduced from 1.0 p.p.m. in the absence of ultrasound to 0.05 p.p.m. in the presence of ultrasound (40 kHz). This effect, coupled with the reported hardness increase of 100 VHN (VHN is the Vickers Hardness Number of the coating), suggests that the use of ultrasound may have important economic advantages, e.g. improved wear resistance, which, providing adhesion is not impaired, would assist the introduction of this new technology into full industrial practice.

References

[1] R. Walker, *Chem.in Britain* (1990) **26**, 251.
[2] B. A. Shenoi, I. L. S.Indira and R. Subramanian, *Metal Finishing* (1970) 57.
[3] S. R.Rich, *Plating* (1955) **42**, 1407.

[4] J. Dereska, E. Yeager and F. Hovorka, *J. Acoust. Soc. Am.* (1957) **29**, 769.
[5] S. R. Rich, *Proc. Amer. Electropl. Soc.* (1955) **42**, 137.
[6] G. O. Mallory, *Trans. of Inst. of Metal Finishing* (1978) **56**, 81.
[7] C. H. de Minjer and A. Brenner, *Plating* (1957) **44**, 1297.

SOLID-SUPPORTED REACTIONS

5.1.15 Sonochemical switching of a reaction pathway — Friedel-Crafts reaction vs nucleophilic displacement

from: T. Ando *and* T. Kimusa, *Shiga University of Medical Science, Shiga 520-21, Japan.*

Introduction

Ando has reported the efficient preparation of benzyl cyanide from benzyl bromide (70%) by direct reaction of the benzyl bromide with KCN in toluene in the presence of activated alumina and ultrasound. This is an interesting reaction in that it is representative of a whole range of supported reactions which are enhanced by ultrasound. The displacement does not occur in the absence of alumina. This is an example of a reaction which follows a completely different course in the absence of ultrasound. Without ultrasound the product formed at the same temperature and under otherwise identical conditions, is a mixture of the 2- and 4-benzyltoluenes produced as a result of a Friedel–Crafts reaction catalysed by the alumina (Scheme 5.14) [1].

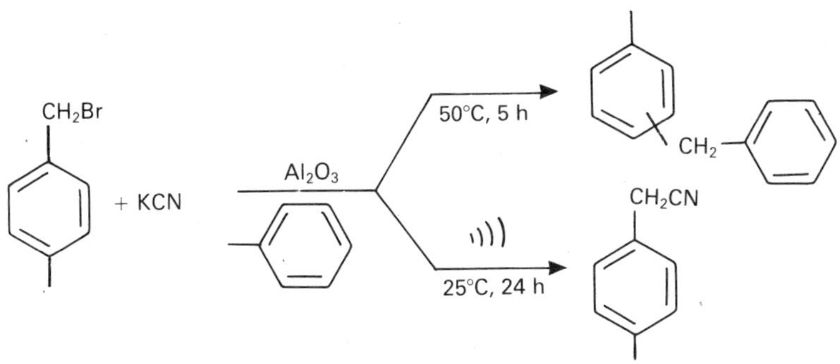

Scheme 5.14

Method

A Pyrex culture tube (25–150 mm, 48 cm^3) fitted with a Teflon-coated screw cap was charged with a mixture of benzyl bromide (0.55 g, 3.2 mmmol), toulene (5 cm^3), powdered KCN (1 g, 16 mmol) and alumina (1.6 g, Merck type 90 for column chromatography, neutral activity 1, 70–230 mesh) and immersed in a constant temperature bath at 50°C for 5 h. After filtration the filtrate afforded a mixture of 2- and 4-benzyltoluene in a total yield of 75%.

In contrast to this result, when the same mixture was immersed in a thermostatted ultrasonic cleaning bath (Branson B-221, 100 W, 45 kHz), in the optimum position, for 24 h benzyl cyanide was obtained in 71% yield.

Discussion

The reaction in the absence of ultrasound parallels some examples of the way in which alumina pretreated with acid has been used as a Friedel–Crafts catalyst [2].

Interestingly the reaction carried out by first pretreating the mixture of KCN benzene and alumina with ultrasound for 3 h followed by conventional mechanical agitation for 24 h gave only 14% of the Friedel–Crafts adducts and no substitution. In fact both the substitution and alkylation reactions were retarded by pre-absorption of KCN. Clearly sonication is essential for the nucleophilic substitution reaction to take place.

In the absence of KCN the mixture of toluene and benzyl bromide gave the Friedel–Crafts adducts smoothly in high yield (*c.* 80%) in 3 h, either with conventional agitation or under sonication. These results indicate that ultrasound assists the contact of KCN with alumina, decreasing the catalytic ability of alumina for a Friedel–Crafts reaction and enhancing nucleophilic attack by CN^- ion at the alumina surface.

References

[1] T. Ando, S. Sumi, T. Kawate, J. Ichihara and T. Hanafusa, *J. C. S. Chem. Commun.* (1984) 439.

[2] *Friedel–Crafts and related reactions*, ed G. A. Olah (1964) Vol. 2, Part 1, Wiley, NY, 425.

5.1.16 Sonochemically enhanced synthesis of α-aminonitriles

from: T. Ando and T. Kimusa, *Shiga University of Medical Science, Shiga 520-21, Japan*.

Introduction

A classical method for the preparation of amino acids is the Strecker synthesis. Recently a new approach to this synthesis has been reported in which amino nitriles have been prepared sonochemically in high yield by the direct reaction between an aldehyde, KCN and NH4Cl in acetonitrile (Scheme 5.15) [1]. This reaction is selective only in the presence of ultrasound and alumina (Table 5.2.).

$$\mathrm{PhCHO + KCN + NH_4Cl \xrightarrow[CH_3CN]{Al_2O_3} PhCH(OH)CN}$$

Scheme 5.15

Table 5.2 — Sonochemical synthesis of aminonitriles

	Reaction Products in CH_3CN at 50°C			
	PhCHO	Ph–CH(OH)–CN	Ph–CH(NH_2)–CN	Ph–CH(NH_2)–CO–Ph
Stirred	35	38	6	21
Stirred+Al_2O_3	8	19	64	9
Sonicated	11	45	23	22
Sonicated+Al_2O_3	0	3	90	7

Method

In a Pyrex culture tube (25–150 mm, 48 cm^3) fitted with a Teflon-coated screw cap, a mixture of KCN (0.63 g, 10 mmol), NH_4Cl (0.59 g, 11 mmol) and alumina (1.5 g, Merck type 90 for column chromatography, neutral activity 1, 70–230 mesh) in CH_3CN (5 cm^3) was initially subjected to ultrasound at 50°C (Branson B-221, 100 W, 45 kHz) for 10 min. After this time benzaldehyde (0.53 g, 5 mmol) was added and sonicated at 50°C for a further 24 h. After filtration of solid substances α-aminoaceto-nitrile was isolated in 90% yield.

Discussion

For this reaction the yield of the target aminonitrile is poor under normal stirred conditions; the predominant products are starting material, benzoin and hydroxyni-trile (Table 5.2). The presence of alumina suspended in acetonitrile increases the proportion of aminonitrile, but the results make it clear that the optimum reaction conditions require that, in addition to suspended alumina, sonication is required. Under these conditions the yield of the aminonitrile reaches 90%.

Reference

[1] T. Hanafusa, J. Ichihara and T. Ashida, *Chem. Letts.* (1987) 687.

5.1.17 A Wittig–Horner reaction using an activated barium hydroxide catalyst (C-200)

from: J. V. Sinisterra Gago, *Departemento de Quimica Organica, Universidad Complutense, 28040 Madrid, Spain.*

Introduction

Ultrasonic irradiation of solid barium hydroxide leads to substantial improvements in yield, rate and reductions in catalyst concentration for a variety of reactions of aromatic aldehydes such as the Cannizzaro reaction, Michael addition, Claisen Schmidt reaction, aldol condensation and the Wittig–Horner reaction.

The reaction chosen to exemplify the use of the barium hydroxide catalyst is the conversion of a benzaldehyde to ethyl cinnamate (Scheme 5.16) [1,2].

Method

The Wittig–Horner reaction was carried out on a small scale in a test tube by mixing benzaldehyde (0.265 g, 2.5 mmol), diethyl phosphonoacetate (0.56 g, 2.5 mmol),

$$PhCHO + (EtO)_2P(O)CH_2COOEt \xrightarrow{Ba(OH)_2} PhCH{=}CH{-}COOEt + (EtO)_2P(O)O$$

Scheme 5.16

THF solvent (3 cm^3) and H_2O (0.05 cm^3) [3] with C-200 catalyst (150 mg). The mixture was sonicated in an ultrasonic cleaning bath (P-Selecta S.A.Model 513, 150 W) for 10 min. The mixture was allowed to stand for 24 h in a refrigerator at 4°C to produce sedimentation of the solid. Similar results are obtained if the sample is centrifuged after sonication. The liquid was analysed by HPLC using a Perkin Elmer Series 2 chromatograph with UV–visible detector (λ=254 nm) and anthracene as internal standard. MeOH/H_2O (80/20 v/v) was used as eluant with a flow rate of 0.8 $cm^3\ min^{-1}$ through a 5-μm C_{18} column [4].

The activated catalyst [5,6] was obtained from commercial barium hydroxide octahydrate (Probus S.A.) by heating at 200°C for 3 h in a nickel crucible and powdering the resultant solid to 250 mesh. Then the solid was stored in a desiccator over NaOH pellets to diminish the rehydration and the adsorption of CO_2. Content of Ba(II)=71∓2% wt. Empirical formula $Ba(OH)_2 \cdot 1.0–1.5\ H_2O$.

Discussion

By means of selective site poisons it was established that the reactions involved reducing sites on the catalyst, and a mechanism involving single electron transfer from the catalyst to the substrate was implicated.

Athough C-200 gives the best yields the process can also be carried out to lower conversions with commercial potassium carbonate ($K_2CO_3 \cdot 1.5H_2O$), potassium hydroxide ($KOH \cdot 3H_2O$) and barium hydroxide ($Ba(OH)_2 \cdot 8H_2O$).

References

[1] A. Fuentes, J. M. Marinas and J. V. Sinisterra, *Tetrahedron Lett.* (1987) **28**, 2951.

[2] J. V. Sinisterra, A. Fuentes and J. M. Marinas, *J. Org. Chem.* (1987) **52**, 3875.

[3] The need for a small quantity of water in this preparation does not seem to be understood, it is not required in the Claisen–Schmidt reaction (section 5.1.18) below.

[4] This procedure gives almost quantitative yields and is generally applicable for aromatic and heterocyclic aldehydes. A variety of solvents can be used, e.g. EtOH (96%), cyclohexane, 1,4-dioxane and *N,N*-dimethylformamide.

[5] A. Garcia Raso, J. Garcia Raso, B. Campaner, R. Mestres and J. V. Sinisterra, *Synthesis* (1982) 1037.

[6] J. Barrios, R. Rojas, A. Alcantara and J. V. Sinisterra, *J. Catal.* (1988) **112**, 528.

5.1.18 A Claisen–Schmidt Condensation using an activated barium hydroxide catalyst (C-200)

from: J. V. Sinisterra Gago, *Departemento de Quimica Organica, Universidad Complutense, 28040 Madrid, Spain.*

Introduction

The synthesis of a range of chalcones can be achieved by the condensation of acetophenone with an aldehyde in the presence of C-200 catalyst (Scheme 5.17) [1].

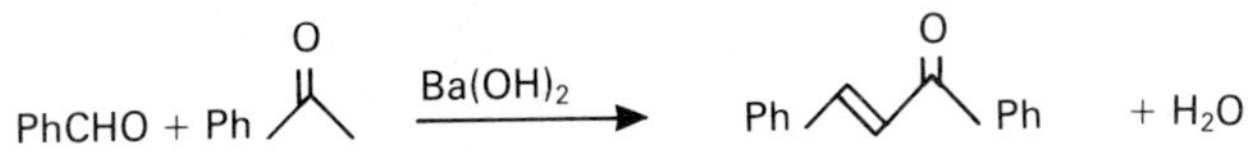

Scheme 5.17

Method

The Claisen–Schmidt condensation was carried out by mixing benzaldehyde (2.1 g, 19.6 mmol) and acetophenone (2.35 g, 19.6 mmol) with THF solvent (10 cm^3) and 10 mg of C-200 catalyst in a 50-cm^3 conical flask. The mixture was sonicated in an ultrasonic cleaning bath (P. Selecta Model 513, 150 W) for 10 min and then left to stand overnight at 4°C to produce sedimentation of the solid. Similar results can be obtained if the sample is centrifuged after the sonication. The samples were analysed by HPLC as described in (section 5.1.16), above and near quantitative yields of *trans*-chalcone were obtained.

Discussion

The synthetic procedure is general for aromatic and heterocyclic aldehydes. As was found in the Wittig–Horner reaction, commercial potassium carbonate, barium hydroxide and potassium hydroxide can be used, but lower yields are obtained than in the case of activated C-200 catalyst.

References

[1] A. Fuentes, J. M. Marinas and J. V. Sinisterra, *Tetrahedron Lett.* (1987) **28**, 4541.

SONOCHEMICALLY ENHANCED KINETICS

The following experiments can be performed at several temperatures and in this way it can be demonstrated that the effect of ultrasound increases as the temperature of the reaction is lowered. As with any kinetics experiment involving ultrasound, there is a problem over obtaining absolute reproducibility but this degree of accuracy is probably unnecessary for the purposes of undergraduate class experiment. It should be noted, however, that accuracy and reproducibility are certainly possible when following sonochemical reactions, although more sophisticated apparatus, such as a cup-horn, is required (see section 5.1.21).

5.1.19 The rate of decomposition of benzene diazonium chloride

from: Coventry Polytechnic Sonochemistry Group

Introduction

This experiment is based upon an undergraduate experiment which we have used for many years in our laboratories. Benzene diazonium chloride is first prepared and then decomposed thermally to nitrogen and phenol (Scheme 5.18). The first order

$$Ph{-}N_2^+Cl^- + H_2O \rightarrow Ph{-}OH + N_2 + HCl$$

Scheme 5.18

decomposition is followed by monitoring the rate of evolution of nitrogen gas, by volume, using a standard gas measuring burette in which the volume of gas liberated can be measured at atmospheric pressure [1].

Method

The diazonium salt was prepared by taking 5 cm^3 water in a boiling tube and adding concentrated HCl (5 cm^3) followed by aniline (1.5 cm^3). The mixture was cooled in ice to below 5°C then 25 cm^3 of cold sodium nitrite solution (of strength 4.4 g in 100 cm^3 water) was added. This solution was made up to 100 cm^3 with cold water, thoroughly mixed, and kept on ice to be used as the stock solution for kinetic studies.

In the case of normal (non-ultrasonic) runs 25 cm^3 of stock solution was placed in a quickfit flask (50 cm^3) equipped with a magnetic stirrer follower and was placed in a thermostatted bath at the required temperature and stirred using a submersible stirrer. (Frequent swirling of the mixture is an inexpensive alternative to mechanical stirring.) The flask was then connected to the gas-measuring burette and, after a minute or two of equilibration, timing was started and readings of gas volume evolved with time were recorded, taking care to make the measurements at atmospheric pressure. The reaction was generally followed for at least one half-life.

Infinity values were obtained, without disconnecting the flask, by warming the reaction mixture to about 70°C for 30 min and taking the reading when the vessel had been cooled back to the temperature used for the kinetic runs.

The ultrasonic runs were performed in the same manner except that:

(a) Ultrasonic agitation eliminated the need for stirring.
(b) The ultrasonic bath temperature was held approximately constant by the addition of ice.
(c) Since the temperature inside the reaction vessel was always a few degrees above that of the bath it was necessary to physically measure it. This could be done, after the infinity value had been measured, with the bath on and at the temperature used for kinetic measurements.
(d) It was noticeable that when the ultrasonic bath was first switched on the solution rapidly degassed; this should be allowed to finish before the kinetic measurements are started.

Discussion

The first order rate constants obtained from these experiments are given in Table 5.3.

Table 5.3 — Kinetic data for the decomposition of phenyl diazonium chloride

Temperature (°C)	Rate ($min^{-1} \times 10^2$)		
	Normal	Sonicated	Enhancement
15	No reaction	0.03	Very large
30	0.77	4.45	6.36
50	10.44	12.83	1.23

The enhancement in reactivity due to ultrasound is clearly seen to be inversely related to the temperature of the system.

References

[1] T. J. Mason, J. P. Lorimer and J. Moorhouse *Education in Chemistry* (1989) **26**, 13.

5.1.20 The rate of solvolysis of 2-chloro-2-methylpropane in aqueous ethanol (using a cleaning bath)

from: Coventry Polytechnic Sonochemistry Group.

Introduction

This experiment is based upon a standard undergraduate experiment in which the first order hydrolysis of a haloalkane is followed by monitoring the change in conductivity of the solution as HCl is liberated (Scheme 5.19).

$$(CH_3)_3CCl + H_2O \rightarrow (CH_3)_3COH + HCl$$

Scheme 5.19

Method

Conductivity water and absolute ethanol may be used without further purification to make up the required solvent compositions by mass. A stock solution of 2-chloro-2-methylpropane was prepared by dissolving 1 cm^3 of freshly distilled material in 9 cm^3 absolute ethanol.

In the case of normal (non-ultrasonic) runs 25 cm^3 of the appropriate solvent mixture was placed in a 50-cm^3 conical flask immersed in a constant temperature bath and allowed to obtain thermal equilibrium. Twenty-five microlitres of the stock substrate solution was injected and the change in conductivity of the reaction was monitored using platinum electrodes. The reaction was stirred throughout using a magnetic stirrer follower and a submersible stirrer. (Frequent swirling of the mixture is an inexpensive alternative to mechanical stirring.) The reaction was generally followed for at least one half-life. First order rate constants are determined with an infinity value or by using the Guggenheim method.

All ultrasonic reactions were performed using precisely the same method, except for the modifications noted in (a) to (d) above (section 5.1.19). Physical measurement of the reaction temperature in the conical flask was made with a thermocouple.

Discussion

Approximate values for the first order rate constants obtained from these experiments are given in Table 5.4. The enhancement in reactivity due to ultrasound is

Table 5.4 — Specimen kinetic data for the hydrolysis of 2-chloro-2-methylpropane in aqueous ethanol

Ethanol in water (%w/w)	Temp. (°C)	Rate constant (no ultrasound) $\times 10^5$	Ultrasonic rate increase
30	10	45	3.0
30	20	160	1.3
50	10	3	12.5
50	20	11	2.5

clearly seen to be inversely related to the temperature of the system. More accurate measurements are possible using a cup-horn system (see section 5.1.21).

Reference

[1] Timothy J. Mason, *Canadian Chemical News* (1991) in press.

5.1.21 The rate of solvolysis of 2-chloro-2-methylpropane in aqueous ethanol (using a cup-horn)

from: Coventry Polytechnic Sonochemistry Group

Introduction

This experiment is essentially a more accurate version of that described above (section 5.1.20). The results obtained using a cup-horn can be reliably compared with those obtained for the rates of solvolysis of 2-chloro-2-methylpropane obtained in a variety of mixed aqueous media using conventional methodology [1,2,3].

Ultrasonic irradiation has been found to accelerate the solvolysis of 2-chloro-2-methylpropane in aqueous ethanol mixtures [4]. The effects are more marked at lower temperatures and higher alcohol compositions with a 20-fold rate acceleration occurring at 10°C and 60% w/w. The kinetic data has been analysed in terms of the variation in solvent structure and vapour pressure. The results lend support to the existence of a region of maximum structuredness in the binary ethanol/water mixture.

Method

Aqueous ethanol mixtures were prepared by mass using distilled water and absolute ethanol. For non-irradiated kinetic runs, 15 cm^3 of the appropriate solvent mixture was placed in a reaction vessel immersed in a constant temperature bath and allowed to obtain thermal equilibrium. 2-Chloro-2-methylpropane (freshly distilled) in absolute ethanol; 1:9 (v/v) was made up as a stock solution for the day's experiments. To start the reaction 15 μl of the stock substrate solution was injected into the reaction vessel, and the change in conductivity of the reaction was monitored, with constant stirring, using platinum electrodes, for a minimum of two-half lives. Constant stirring was maintained throughout the reaction. All ultrasonic reactions were performed using precisely the same method except that the constant temperature bath was replaced by a cup-horn reactor (Chapter 2) operating at 20 kHz and delivering an acoustic intensity of approximately 1 W cm^{-2}. In these cases the appropriate solvent mixture was allowed to attain thermal equilibrium *during sonication* before the substrate was injected.

Discussion

The observed rate constants in the absence of (k_{non}) and presence (k_{ult}) of ultrasound were determined as a function of temperature for various solvent compositions (30, 40, 50% and 60% w/w ethanol/water). When k_{ult} is plotted against reaction temperature at these solvent compositions some interesting observations can be made. At 20 and 30% (w/w) ethanol the curve is similar in shape to that obtained in the absence of ultrasound, i.e. the rate constants increase rapidly with increase in temperature. At 40% (w/w), however, the rate constant increase with temperature is not nearly so large (Fig. 5.3). At 50% and 60% solvent composition a dramatic reversal of the expected Arrhenious behaviour is seen. Below 20°C the rate constant does not fall as expected, but rises rapidly with the lowering of the temperature, indicating that the effect of ultrasound is becoming more marked as the temperature is lowered [4].

The rate enhancements induced by ultrasonic irradiation can be represented as the ratio of rate constants (k_{ult}/k_{non}). Using this ratio, the solvolysis rate in 20% EtOH at 10°C in the presence of ultrasound is twice that in the absence of ultrasound, whereas in 40 and 60% EtOH the rate increases are six- and twentyfold respectively, showing a considerable solvent effect. At 25°C, however, the ratio in all solvents is quite small, with the largest value (1.7) being obtained at a solvent composition of 60%. At 20°C the curve obtained reveals a particularly interesting feature, in that at 50% composition there is a clear maximum in sonochemical effect (Fig. 5.4). It is interesting to note that when sound absorption (attenuation) is monitored against solvent composition in water/ethanol mixtures there is a maximum in the sound absorption composition curve which is also at about 50% w/w [5].

In order to explain this we must consider the structure of aqueous ethanol solvents. In simple terms we can envisage water as a three-dimensional network, full

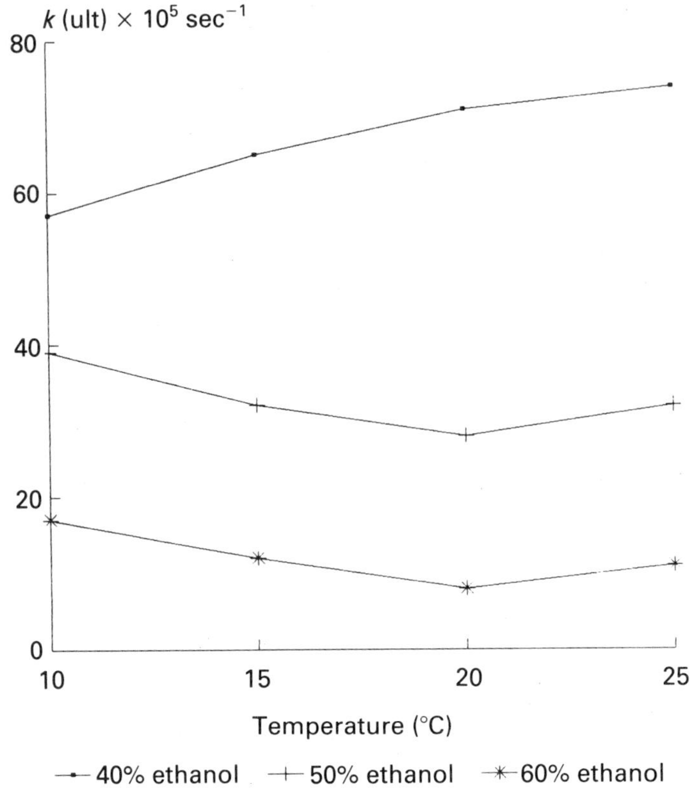

Fig. 5.3 — Ultrasonic rate constants for *t*-butyl chloride hydrolysis.

of holes, in equilibrium with free water. The introduction of small amounts of ethanol to water is accompanied by the filling of these holes to produce a stronger structure and an increase in hydrogen bonding. Obviously at some composition here 50% w/w, the alcohol can no longer be accommodated by the water and the structure begins to break down, producing more 'free' water. The explanation offered for the existence of this maximum in the sound absorption curve therefore is that the passage of the sound wave through the medium causes breaking of the hydrogen-bonding structure. The greater the structure the more there is to break and the greater the absorption. Hence if water is an equilibrium mixture of structured and free water, the application of ultrasound to the medium will drive the equilibrium in favour of 'free' water. The medium is apparently 'more aqueous' and will thus provide an increase in the solvolysis rate.

References

[1] E. Grunwald and S. Winstein, *J. Am. Chem. Soc.* (1948) **60**, 846.
[2] A. H. Fainberg and S. Winstein, *J. Am. Chem. Soc.* (1956) **68**, 2760.
[3] R. E. Robertson and S. E. Sugamori, *J. Am. Chem. Soc.* (1969) **91** 6254.

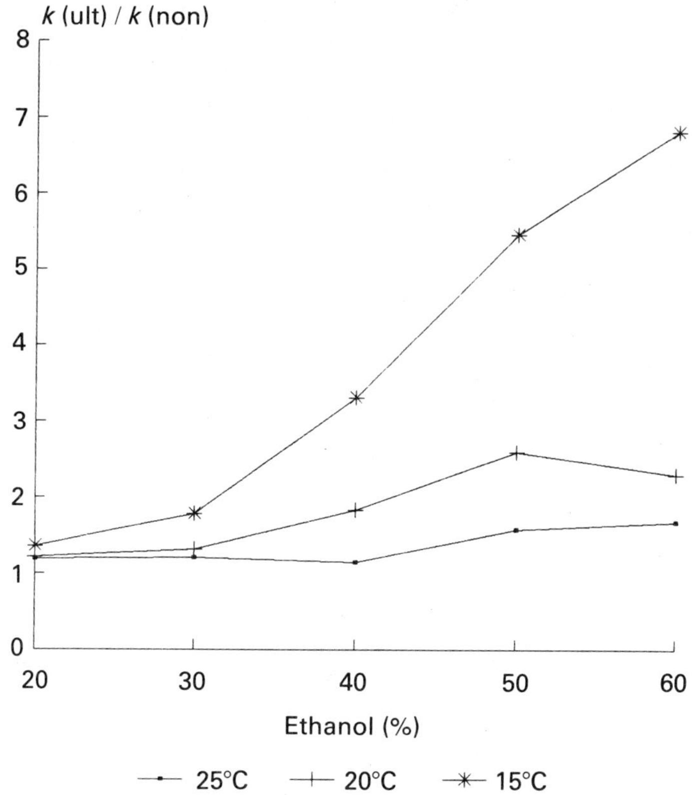

Fig. 5.4 — Ultrasonic rate enhancements for *t*-butyl chloride hydrolysis.

[4] T. J. Mason, J. P. Lorimer and B. P. Mistry, *Tetrahedron* (1985) **41**, 5201.
[5] J. C. Burton, *J. Acoust. Soc. Am.* (1948) **20**, 186.

MISCELLANEOUS EXPERIMENTS USING AN ULTRASONIC BATH

5.1.22 Preparation of 1,2,5,6-diisopropylidene glucofuranose

from: C. Einhorn, A. De Nicola *and* J. L. Luche, *L.E.D.S.S., Chimie Recherche, Université Scientifique et Médical de Grenoble, France.*

Introduction

Sugars hold a special interest for the organic chemist. They constitute a vast series of compounds with important biological properties either in a natural or modified form, and some of them are among the more common and inexpensive sources of chiral compounds [1]. The chemistry of sugars has its own peculiarities in that these compounds are insoluble in organic media, while they are highly soluble in the (frequently less desirable) aqueous phase. Also, because of the presence in the molecule of several hydroxy groups, the first step of a chemical transformation from these compounds is frequently the protection of an alcohol functionality, which makes the compounds more soluble in organic solvents, and allows further selective reactions. Among many possibilities, one of the more convenient is the formation of ketals by condensation of the sugar with a carbonyl compound, with acetone, benzaldehyde and cyclohexanone being the more common [2]. With acetone, for instance, the reagent can be used as the reaction medium, and the heterogeneous process is then conducted in the presence of a strong acid (protic or Lewis). Because the system is heterogeneous, the sugar being insoluble in acetone, a rather long contact time is necessary, and during this period side reactions can take place. It would clearly be of great use if the reaction rate could be increased while retaining good yields without loss of product purity. The application of sonochemical methodology was seen as a means of obtaining such improvements [3]. The validity of this prediction has been tested in the case of the preparation of 1,2,5,6-diisopropylideneglucofuranose (Scheme 5.20).

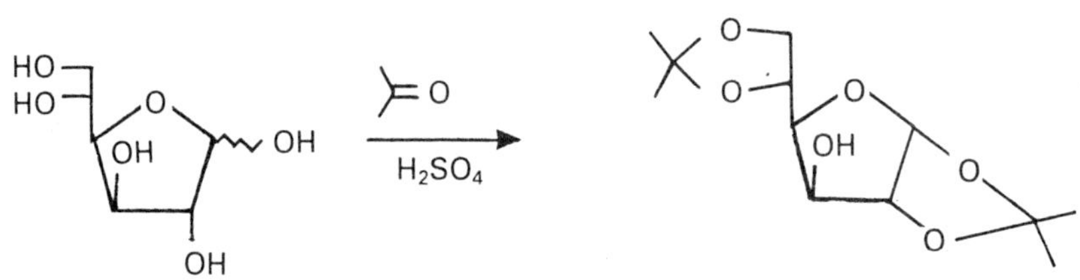

Scheme 5.20

Method

A laboratory cleaning bath (Kerry Ultrasonics, Pulsatron 250, 38 kHz) filled with water was thermostatted to 18–20°C by running tap water in a stainless steel coil. Zones of maximum energy emission in the bath were detected by the disturbance of the surface of the water, which was adjusted to give a height of 8 cm (i.e. 2 wavelengths). The reaction flask was centred on a maximum of energy, the exact position corresponding to the formation of mist above the surface of the liquid inside the vessel. It was then firmly clamped.

The flask (round-bottomed, 2-l, Pyrex) was filled with acetone (1500 cm^3, Aldrich, HPLC-grade) and commercial 98% sulfuric acid (40 cm^3). Then 50 g of powdered glucose was added, all at once, under sonication. A mechanical stirrer was operated at a slow rate to prevent sedimentation (50–60 r.p.m.) and sonication was continued for 1 h. At the end of this period, the solid was almost completely dissolved and the liquid phase was light yellow with a slight turbidity. The reaction was then completed by sonication for a further 15 min (total time 75 min). The mixture was then rapidly poured into a solution of ammonia (120 cm^3, 20%) [4] cooled in ice water. The solid ammonium sulfate was filtered off and washed with acetone (two portions of 100 cm^3). The colourless solution was evaporated at a temperature below 30–35°C and then the solid was dissolved in hot heptane. In some cases a turbidity appears, which can be separated by decanting the supernatant liquid [5]. After the mixture had been cooled, the crystals were separated and dried in air to yield 53 g of 1,2,5,6-diisopropyliidene glucofuranose as a white solid (73%). m.p. 105–106°C (literature value 105–109°C) [6], ($[\alpha]_D$ −18.7 acetone) (literature value −19, acetone) [6]; the IR and NMR spectra were consistent with the structure [7].

Discussion

This reaction constitutes an example of a sonochemical process based on the mechanical effects of ultrasound. The macroscopic effect consists of the dispersion of the solid in the reaction medium. It has been observed that, under the conditions specified above, a limit exists for this effect at *c*. 30 g of glucose. The power of the generator is insufficient to disperse larger quantities and mechanical stirring is necessary to prevent sedimentation. The microscopic effect of ultrasound, on the other hand, seems to be unaffected by the size of the sample, as the reaction time undergoes only minor variations when the quantity of starting material is increased from 5 g (60 min) to 50 g (75 min). The advantages of the sonochemical method are obvious by comparison with the same reaction performed under conventional conditions, which is reported to give a 42% yield after 5–8 h stirring [6].

The method was tested with success on the gram scale in several other cases, as shown in Table 5.5.

Table 5.5 — Sugar ketal formation assisted by sonication

Starting material	Carbonyl compound	Time (mins)	Yield (%)
D-Glucose	Cyclohexanone	45	75
D-Galactose	Acetone	75	72
D-Mannose	Acetone	15	86
D-Mannose	Cyclohexanone	60	43
D-Mannitol [7]	Acetone	15	80

It should be noted that, in the case of the preparation of the dicyclohexylidene derivatives, only 2 equivalents of cyclohexanone are necessary (2.2 g cyclohexanone for 1 g of sugar), and 30 cm^3 of ethyl acetate or acetonitrile are used as diluent [3]. For mannitol, the triisopropylidene derivative is obtained in excellent yield.

References

[1] S. Hanessian, *Total synthesis of natural products: the chiron approach*, Pergamon Press, Oxford, 1983.
[2] J. Gelas, *Adv. Carbohydr. Chem. Biochem.*, (1981) **39**, 71.
[3] C. Einhorn and J. L. Luche, *Carbohydr. Res.* (1986) **155**, 258.

[4] It is preferable to use the stoichiometric amount of ammonia for this neutralization. It was observed that the yields are decreased by excess base.
[5] At this stage, it is possible to separate *c*. 5% of 1,2-isopropylideneglucofuranose.
[6] O. T. Schmidt, *Methods Carbohydr. Chem.* (1963) **2**, 318.
[7] IR: 3500–3300, 2995, 1425, 1370 cm^{-1}, NMR (80 MHz, $CDCl_3$, TMS internal reference): 5.9(d, 1H), 4.5(d, 1H), 4.4–3.9(m, 5H), 2.5(d, 1H), 1.6–1.3(4s, 12H) p.p.m.

5.1.23 The generation and spin trapping of free radicals

from: P. Riesz, *Radiation Oncology Branch, National Cancer Institute, Bethesda, Maryland 20892, U.S.A.*

Introduction

The identification of free radical intermediates generated during sonolysis is important in establishing sonochemical mechanisms. These short-lived species can often be identified by spin trapping and subsequent ESR examination of the spin adduct formed (Chapter 1) [1].

Method

The aqueous solution (3 cm^3) to be investigated, containing the appropriate spin trap (commonly 1–5 mg cm^{-3}) was placed in a Pyrex test tube, fitted with a rubber septum, fixed by means of a lucite holder in the centre of the sonication bath (Bransonic 1200, frequency 48 kHz). The tube was immersed to a depth where the level of the liquid inside was the same as that of the bath water outside in order to obtain sufficient reproducibility in the standing wave field. The temperature of the coupling water was 25°C. Argon, or another appropriate gas, was bubbled through the solution via a fine needle attached to a Teflon tube at a fixed flow rate (normally 40 cm^3/min) for 10 min before sonication and also during sonication. After such sonolysis experiments, the change of absorbance of a Fricke dosimeter solution (1 mM ferrous ammonium sulfate, 1 mM sodium chloride in 0.4-M sulphuric acid) at 304 nm was measured to be 0.61 ∓ 0.06 (SD, $n=5$) for argon and 0.47 ∓ 0.05 (SD, $n=5$) for oxygen, respectively.

Immediately after sonolysis, the 300 μl of solution was transferred into a standard ESR quartz flat cell (60×10×0.25 mm). A Varian E-9, X-band spectrometer was used for ESR measurements and the magnetic field was 100 kHz.

References

[1] P. Riesz, T. Kondo and C. M. Krishna, *Ultrasonics* (1990) **28**, 295.

5.1.24 The observation of simultaneous reactions in an ultrasonic field — polymer degradation versus iodide oxidation

from: A. Henglein, *Hahn-Meitner-Institut, Bereich Strahlenchemie, 1000 Berlin 39.*

Introduction

The following experiment employs the specially designed 1 MHz baths of the Henglein group described in Chapter 2.

The unusual feature of ultrasonically induced polymer degradation is that it occurs even during sonication in the more volatile solvents, such as benzene and carbon disulfide, where cavitational collapse must be 'soft'. This suggests that the hydrodynamic shear forces in the vicinity of the collapsing bubbles are not necessarily greatest under conditions where the bubble temperature is greatest. It is also observed that the rate of degradation does not decrease steeply at high intensities, which is a feature of redox reactions such as the sonochemical oxidation of aqueous iodide to iodine.

An interesting comparison can be made between the two types of sonochemical reaction (redox and degradation) when both types are run concurrently [1].

Method

The aqueous solutions used for the experiments were KI (0.2-M) and poly(acrylamide) (5×10^{-2}, EGA-Chemie, average molecular weight 5×10^{6}). Power input to the system was assessed using the calorimetric technique.

Separate measurement of the rate of degradation of the polymer solution at fixed power was achieved by measuring the change in viscosity of the aqueous solution with sonication time, using the formula:

$$n = n_0[1/(1 + k_d^0 t)]$$

where n_0 is the relative viscosity before irradiation, n is the relative viscosity at time t and k_d^0 is the degradation constant for continuous irradiation. A plot of n_0/n versus irradiation time gives a straight line from whose slope k_d^0 can be calculated. When a series of degradation rate measurements are made at different powers the resulting plot shows that the rate of degradation increases with intensity, as expected, and then reaches a broad maximum until a slight decrease occurs at very high intensity (Fig. 5.5).

The liberation of iodine from KI is followed spectrophotometrically at 350 nm as described in Chapter 1. In this case a plot of rate of liberation of iodine versus power reveals a steep initial rise in rate as power increases, followed by a rapid drop at higher powers (Fig. 5.6).

A series of experiments were then performed in which a solution containing both KI and poly(acrylamide) was subjected to sonication.

Discussion

A plot of the ratio of the rate of polymer degradation to that of iodide oxidation versus intensity gives the curve in Fig. 5.7. It rises steeply in the high intensity range where the I^- oxidation rapidly slows down. This is because the coalescence of bubbles at high intensity leads to larger bubbles in which the temperatures are not high enough to produce many radicals by pyrolysis of water vapour in the bubbles. However, the hydrodynamic shear forces in the vicinity of these larger bubbles are still strong enough to ensure efficient degradation of the polymer.

The ratio of rates was also investigated using 20-kHz horn sonication. It was found that the ratio of polymer degradation to iodide oxidation was much greater at

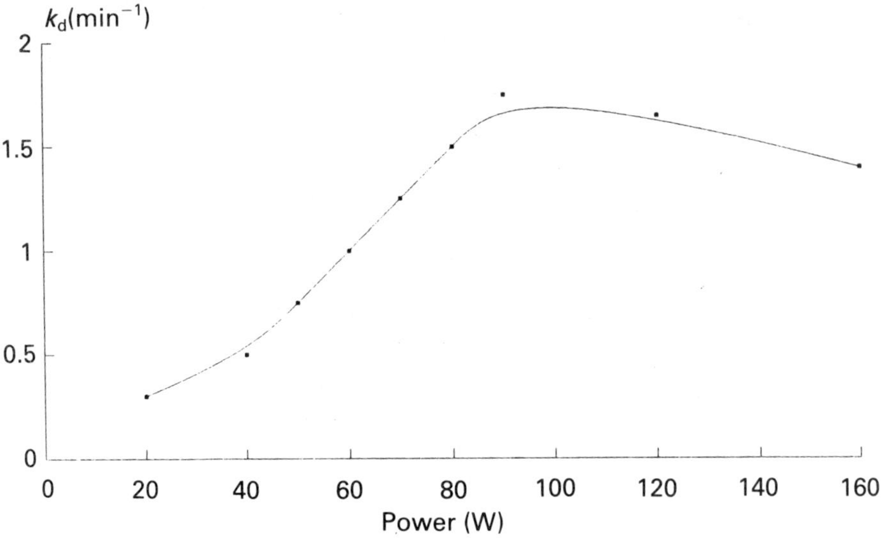

Fig. 5.5 — Degradation of poly(acrylamide) as a function of intensity.

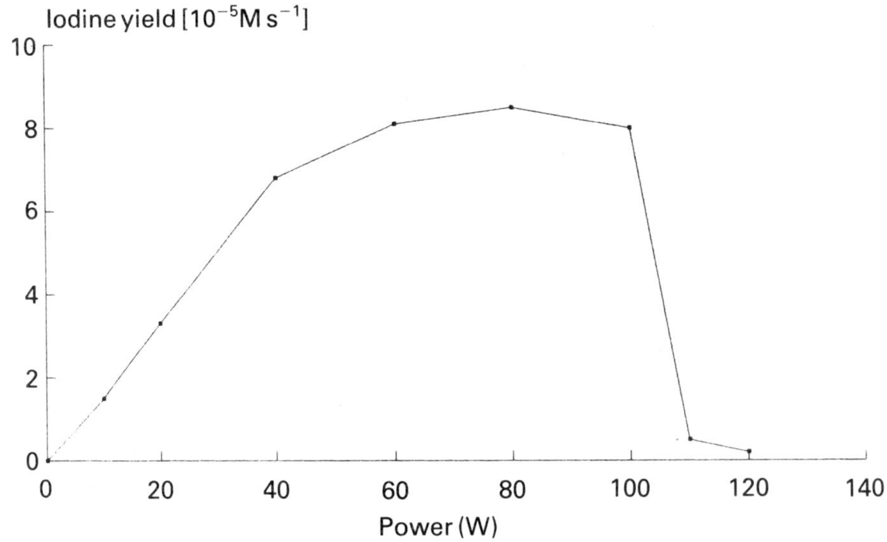

Fig. 5.6 — Iodine yield as a function of intensity.

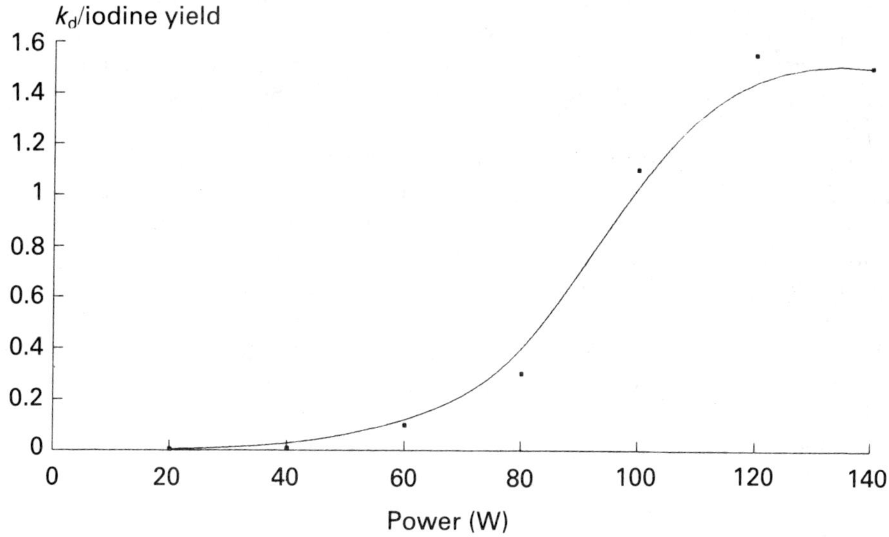

Fig. 5.7 — Ratio of polymer degradation to iodine yield.

all intensities (2 to 12 W cm^{-2}) than at the highest intensity used in the 1 MHz bath experiments. In other words under the probe sonication conditions the mechanical effect is even more pronounced compared with free radical reactions. These conditions were comparable to those which prevail when probe systems are used for biological cell rupture and seems to support the idea that under cell disruption conditions free radical attack on the liberated contents would be far less than an equivalent degradation under higher frequency conditions.

Reference

[1] A. Henglein and M. Guitierrez, *J. Phys. Chem.* (1990) **94**, 5169.

5.2 EXPERIMENTS USING ULTRASONIC PROBES

POLYMER SONOCHEMISTRY

5.2.1 The effect of power ultrasound on the degradation of dextran in aqueous solution.

from: The Coventry Polytechnic Sonochemistry Group.

Introduction

Dextrans are polysaccharides of variable molecular weight consisting of predominantly straight-chain polymers of glucose, in which the linkages between the glucose units are almost entirely of the α-1,6 type.

Although dextrans have been used in a number of industrial applications, the greatest single use has been in the medical field, where partially hydrolysed dextrans have been used as a substitute for blood plasma [1]. Dextran can be complexed with iron to provide a source of the metal for the production of haemoglobin, particularly in veterinary medicine.

For use in medical fields the dextran should not have a very high molecular weight, and so degradation is an important part of commercial production. Industrially the lower molecular weight dextrans are obtained by partial acid hydrolysis of the high molecular weight material generated by fermentation. This process can often take in excess of 24 h and so any method of accelerating the process is of interest to industry.

It has been known for some time that long chain molecules are broken down by ultrasonic waves [2,3]. Although the exact mechanism by which this occurs is open to question, it is generally agreed that it is the hydrodynamic forces which are of primary importance. It is also believed that ultrasonic degradation, unlike chemical or thermal decomposition, is a non-random process, with cleavage taking place at roughly the centre of the molecule and with larger macromolecules degrading the fastest. The consequence of this is that the larger molecules are preferentially degraded. It is also known that there is a limiting molecular weight below which degradation does not take place. This limiting molecular weight has the added effect of narrowing the molecular weight distribution.

Basedow and Ebert [4] investigated the effect of ultrasound on the degradation of dextran in several solvents and found that the solvent characteristics have a considerable effect on the degradation rates and that there is a relationship between the rate constants and the enthalpy of vaporization of the solvent.

Method

The dextran sample used in this study had a relative molecular mass (RMM) of 1 321 000 (supplied by Fisons plc, Holmes Chapel, UK) and was employed as a 2% w/v aqueous solution. A sample of this solution (200 cm^3) was irradiated using a Sonic Systems VC600 ultrasonic generator equipped with a 0.5-inch (12.5 mm)- diameter horn and extender. Reactions were performed in a Rosett cell in a thermostatted water bath at 30°C using 60 W power. The solution was irradiated for 3 h and samples withdrawn at 15, 30, 60, 120 and 180 min.

The original 2% solution, together with all of the samples obtained during sonication, were analysed using gel permeation chromatography (GPC) in terms of mass% versus RMM (i.e. the distribution of molar masses directly from the GPC trace). The degradation data (Fig. 5.8) shows how the molar mass distributions of the sonicated dextran sample change with time. In order to confirm that ultrasonic degradation did not take place by thermal action alone, 200 cm^3 of the 2% aqueous dextran was refluxed for 64 h and also analysed by GPC. This result is shown also in the Figure.

Discussion

A comparison of the thermal data with the sonicated data indicates that even after 6 h of reflux there is still a considerable amount of high molecular weight material left,

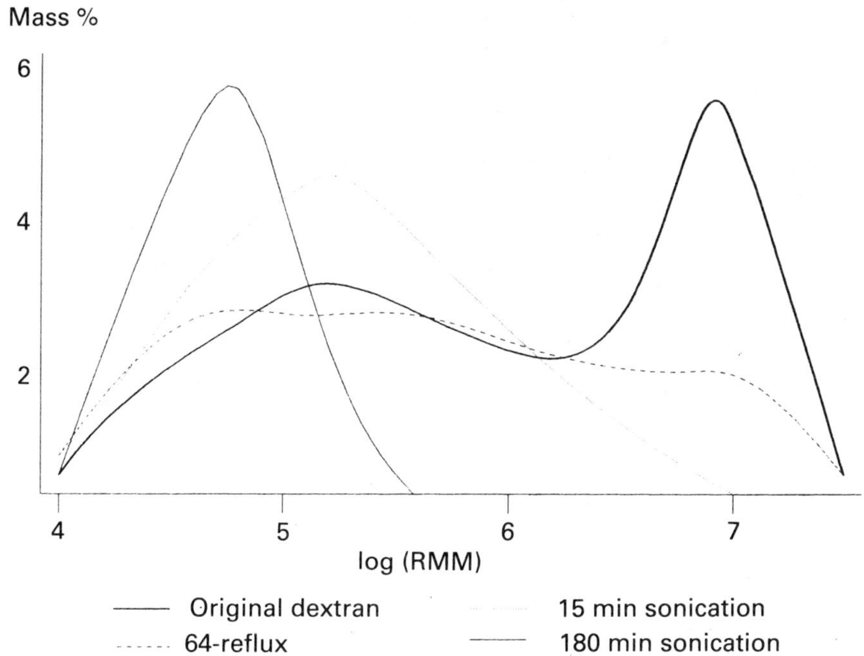

Fig. 5.8 — Dextran degradation.

whereas with the sonicated run most of the high molecular weight material has been removed after only 15 min sonication. Fig. 5.6 also shows conclusively that degradation using ultrasonic irradiation causes considerable narrowing of the molecular weight distribution so that a single peak is obtained after 3 h sonication.

References

[1] C. R. Ricketts, L. Lorenz and W. d'A. Maycock, *Nature* (1950) **165**, 770.
[2] A. Szalay, *Z. Physik. Chem.* (1933) **A164**, 234.
[3] E. Thieme, *Physik. Z.* (1938) **39**, 384.
[4] A. M.Basedow, K. H. Ebert, *Makromol. Chem.* (1975) **176**, 745.

5.2.2 Ultrasonically induced polymerization of methyl methacrylate

from: P. Kruus, *Department of Chemistry, Carleton University, Ottawa, Canada.*

Introduction

The use of ultrasound to modify polymer properties has been of interest for many years, although earlier work tended to concentrate on degradation studies. The effects of ultrasound on radical polymerization processes is now receiving attention, since it offers the possibility of reducing the proportion of radical initiator needed and producing polymers which may differ in properties from conventionally prepared materials.

Method

The reaction was performed in a specially made glass cell (Fig. 5.9). This was

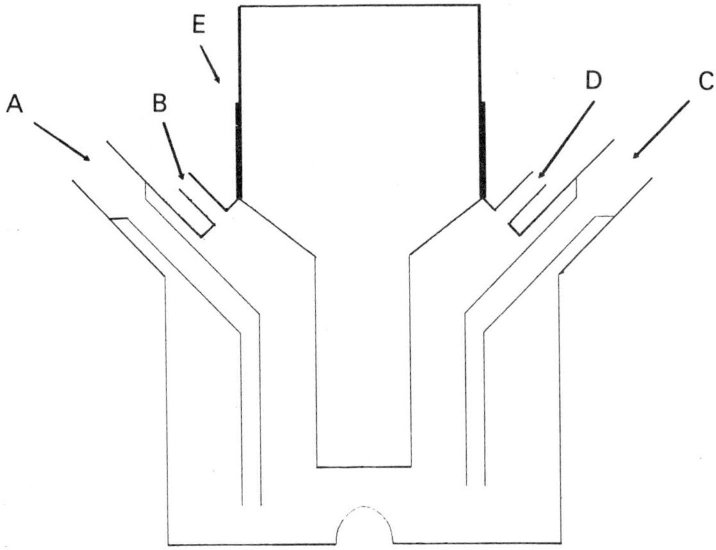

Fig. 5.9 — Polymerization cell.

equipped with a thermocouple well for temperature measurement (A), a port for sample withdrawal (B), a gas inlet for argon gas (C) and a port for monomer introduction (D). Ultrasound was introduced directly into monomer through the tip of a sonic horn (Bransonic W350 Sonifiier, 20 kHz) protruding through a large central opening in the vessel (E). A polyethylene sleeve provided a seal between horn and vessel. An indentation at the base of the vessel located directly below the probe tip resulted in efficient stirring as a result of the strong acoustic streaming. The horn was reproducibly placed in this vessel to within ∓0.5 mm in the horizontal and 0.5 cm in the vertical.

Linde prepurified argon was further purified prior to its use by passing it through an oxygen scavenger (Ridox, Fisher R-30). Activation of the scavenger was effected by reduction with a mixture of hydrogen and argon (5 : 95 by volume) at 350°C for 2 h. The monomer, reagent-grade methyl methacrylate (MMA) (Fisher 0-3629), was purified as follows. Approximately 1 l MMA was washed twice with an equal volume of 10% aqueous NaOH in order to remove the hydroquinone inhibitor. The MMA was then washed twice with equal volumes of distilled water. A slight emulsion resulting from the water washing was removed with a single washing with an equal volume of saturated NaCl. The MMA contains a small amount of water which was removed by standing firstly over Na_2SO_4 (300 g) for 40 minutes then over activated molecular sieves (130 g, Type 4A, Fisher M-514). Reproducible results were only obtained if the MMA dried in this manner was then degassed by boiling it under a vacuum with vigorous stirring for about 10 min immediately prior to distillation. The

MMA was distilled under argon from Cu(II)Cl (*c*. 2 g). About 400 cm^3 of MMA was collected in a special receiving cell which allowed transfer of the monomer into the sonication cell without exposure to the atmosphere. The refractive index of the purified MMA was in agreement with that reported for pure MMA (1.4147 at 298 K). The methanol used was HPLC-grade (Fisher A-452).

Freshly distilled MMA was introduced into the preweighed argon-filled sonication cell directly from the distillation receiving cell without atmospheric exposure. Argon was vigorously bubbled through the monomer for 45 min to ensure saturation. The argon flow rate was reduced to approximately 20 $cm^3\ min^{-1}$ and the cell was placed in a thermostatted bath. The temperature in the cell rose roughly 5°C within the first 5 min after the ultrasound was turned on and could be maintained to ∓1°C thereafter. Measurement of the ultrasonic power input to the liquid monomer (by calorimetry) revealed the range available to be between 0 and 76 W, corresponding to an intensity range of 0 to 27 W cm^{-2}. Reproducibility at any given setting was 3% of a mean over at least three separate determinations.

Samples of MMA (approximately 8 g) were periodically withdrawn with a syringe, weighed, and shot into a 20-fold excess of methanol to precipitate the polymer. The polymer was then collected on preweighed, fine-porosity sintered glass filters, dried on the filter for 18 h and weighed.

Fig. 5.10 shows the percentage conversion of monomer plotted against irradiation time at an acoustic intensity of 20.6∓0.3 W cm^{-2} at three different temperatures.

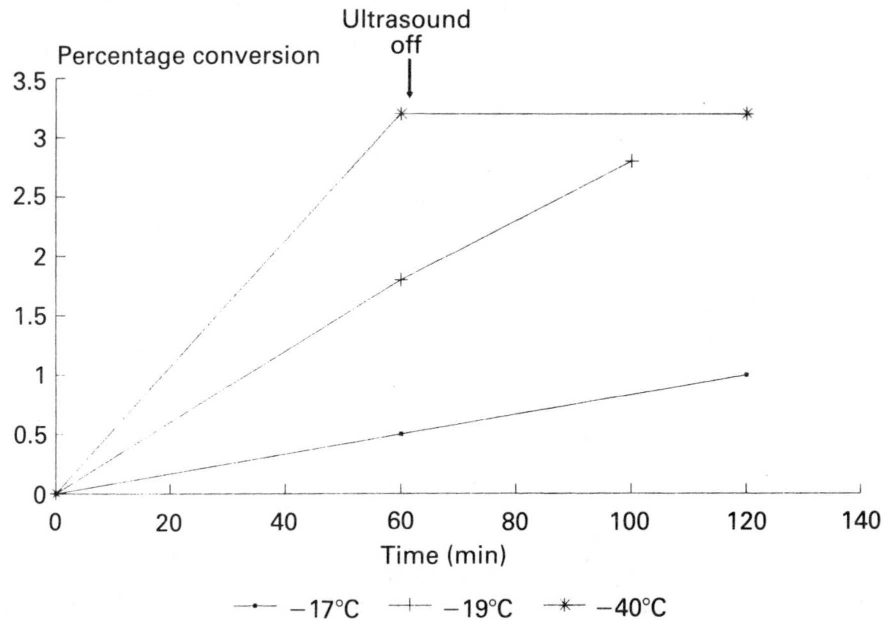

Fig. 5.10 — Monomer conversion against insonation time.

Discussion

The rate of conversion (consumption) of monomer showed an inverse relationship to the temperature, as might be expected if the polymerization depends upon cavitational collapse. At lower temperatures the polymerization was faster because a greater collapse energy was involved at the lower vapour pressure. The number average molecular weight of the polymer obtained by this method was in agreement with values obtained from conventionally initiated bulk polymerization reactions.

References

[1] P. Kruus, *Advances in Sonochemistry* (1991) Vol 2, Ed T. J. Mason, JAI Press.
[2] P. Kruus and T. J. Patraboy, *J. Phys Chem.* (1985) **89**, 3379.

5.2.3 The effect of ultrasound on the encapsulation of titanium dioxide powder by a polymer.

from: The Coventry Polytechnic Sonochemistry Group.

Introduction

Over the past 25 years there has been a general interest in the development of technologies for the encapsulation of fine inorganic powders with organic polymers [1–4]. The general aim of encapsulation is to affect the physical properties of such powders, particularly in terms of increasing their dispersibility in solvents or in composite phases. For example, if the end use of the powder is to be in either the coating field or the production of speciality films, then the two factors which dictate optimum physical properties are firstly an even and small particle size of the original powder, and secondly a uniform coating of each and every particle. This experiment shows how power ultrasound may be used to achieve these effects in an otherwise conventional emulsion polymerization process [5].

Method

The chemical materials employed are methyl methacrylate (MMA), methacrylic acid (MA), ethylene glycol dimethacrylate (EDMA), potassium persulfate (KP), sodium metabisulfite (NaM), aluminium nitrate (AN) and the surfactant Triton-X-100 (TX), which were used without further purification. The water was deionized. The MA and AN are employed as coupling agents to aid in bonding the monomer to the surface of the TiO_2 particles. EDMA is used as a crosslinking agent to improve the stability of the polymer coatings.

The polymerization experiment was carried out, under conventional conditions, in a five-neck kettle flask (1 l) fitted with a stirrer and nitrogen supply line. The powder (TiO_2, 188 g), water (700 cm^3) and MA (1.9 g) were added to the flask and the contents vigorously stirred under nitrogen for approximately 15 min after which time AN (2.8 g) was added. After a further 10 min, EDMA (0.53 g) and 1% solutions of KP (0.019 g) and NaM (0.01 g) are added and the temperature raised to 40°C (*c.* 15 minutes). MA (7.5 g), together with the surfactant (0.38 g), are pre-emulsified with water (50 cm^3) by immersing in an ultrasonic bath. The resulting emulsion was added over a 15-minute period, with stirring, to the reaction flask at 40°C. Further charges of EDMA (0.4 g), KP (0.072 g) and NaM (0.036 g) were added and the temperature raised to 70°C (over about 60 min). The polymerization was left for approximately 4 h under a nitrogen sparge before filtering and washing the product with water and drying overnight at 70°C.

In an ultrasonic run the source of ultrasound is a Heat Systems W225 sonicator (20 kHz) with a horn of 0.5 inch (12-mm) end diameter. The power delivered to the system was nominally 28 W cm^{-2} and was most effective when used only during the mixing period, i.e. it was switched off prior to the 4-h (conventional) polymerization period.

Discussion

The polymer-coated particles were examined using a transmission electron microscope to investigate the nature of encapulsation. Whereas in the absence of ultrasound the standard recipe gave limited, irreproducible coverage (Fig. 5.11(a)), the influence of ultrasonic irradiation showed a marked improvement in the ability of the polymer to coat the particles (Fig. 5.11(b)). In the micrographs the black and translucent parts represent TiO_2 powder and polyMMA polymer coating respectively.

The effect of ultrasound on the particle size of the powder in aqueous suspension was also investigated. A small sample of TiO_2 powder was dispersed, by stirring, in HPLC-grade water in a glass cuvette (1×1×2.5 cm). The volume mean diameter (v.m.d.) of this sample was measured by a laser scattering technique using a Gallai particle size analyser operating on CIS-1 software. The affect of low power (0.5 W cm^{-2}) ultrasonic irradiation on this suspension was assessed by periodic immersion of the cuvette in an ultrasonic bath (Kerry Pulsatron 55, 35 kHz). Initially the powder sample consisted of agglomerates (*c.* 19 μm) which were rapidly broken apart (in less than 30 s) to provide a limiting size of 1.6 μm. Further, the sample showed no sign of agglomeration even after being allowed to stand for a period of 24 h. Thus in the encapsulation procedure involving ultrasonic agitation in the mixing phase, the suspended TiO2 is almost certainly deagglomerated within the first minute, thereby increasing the possibility of single-particle encapsulation.

References

[1] H. G. Decking, *J. Appl. Polym. Sci.* (1965) **9**, 1641; H. G. Decking *J. Appl. Polym. Sci.* (1967) **11**, 23.

[2] A. Blustein, *J. Polym. Sci. A* (1965) **3**, 2653.

[3] K. Nollen, V. Kaden and K. Hammann, *Angew. Makromol. Chem.* (1969) **6**, 1.

[4] K. Fukano and E. Kageyama, *J. Polym. Sci., Polym. Chem. Ed.* (1975) **13**, 1309.

[5] J. P. Lorimer, T. J. Mason, D. Kershaw, I. Livsey and R. Templeton-Knight, *Colloid and Polymer Science* (1991) in press.

EXPERIMENTS INVOLVING METAL POWDERS

5.2.4 Enhancement of an Ullmann coupling reaction induced by ultrasound

from: The Coventry Polytechnic Sonochemistry Group.

The Ullmann coupling of arylhalides to yield biaryls is widely used in aromatic chemistry and normally is carried out by heating the halide (with or without solvent)

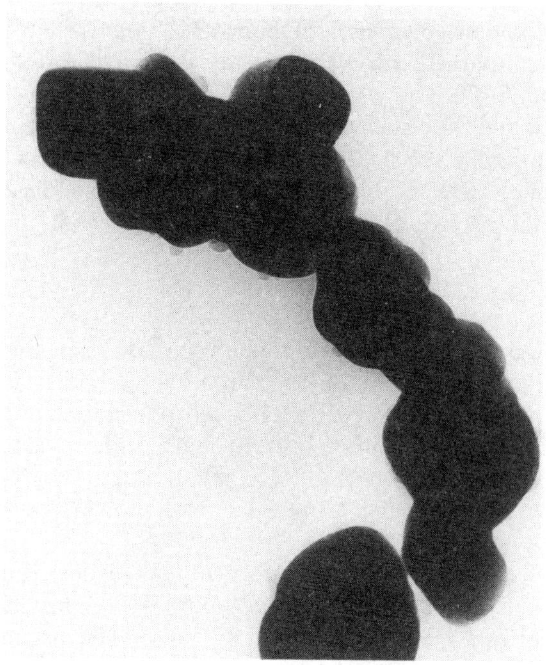

(a)

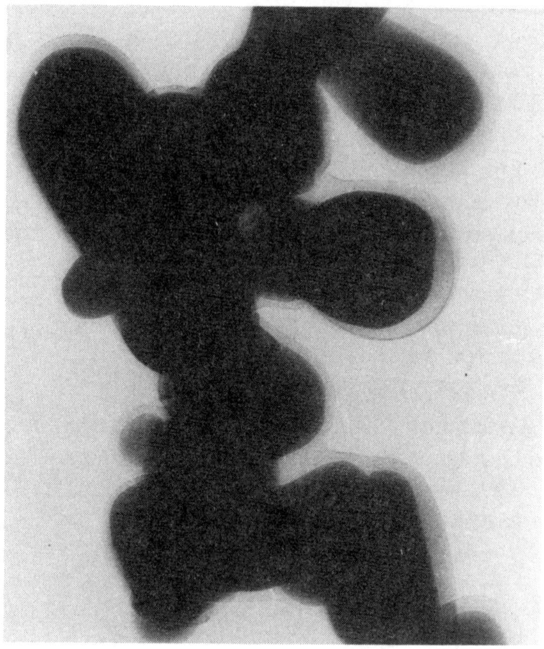

(b)

Fig. 5.11 — Effect of sonication on encapsulation.

in the presence of copper powder [1,2]. Classical Ullmann conditions involve high temperatures and extended reaction times. An Ullmann reaction which takes place under less forcing conditions is the conversion of 2-iodonitrobenzene to 2,2-dinitrobiphenyl (99.6%) over a period of 48 h at 60°C using a tenfold excess of copper (Scheme 5.21) [3]. This was the system chosen for the experiment below [4].

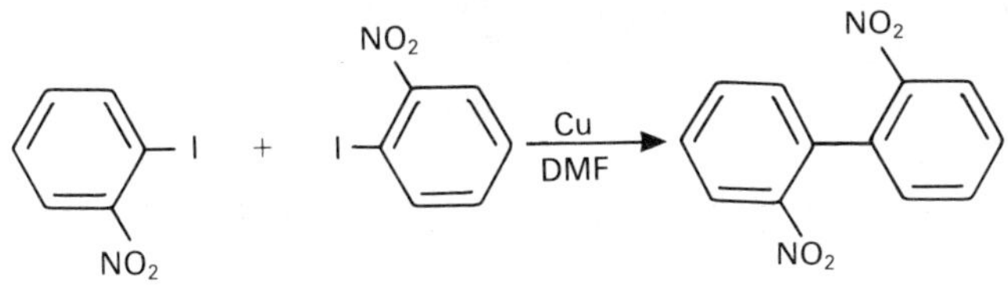

Scheme 5.21

Method

The reactions were performed in a Rosett cell (capacity 50 cm^3), immersed in a bath thermostatted at 60°C. In a typical experiment copper bronze (3.8 g, 59 mmol) and 2-iodonitrobenzene (3.7 g, 15 mmol) were reacted in dry dimethylformamide DMF (30 cm^3) as solvent. The reaction temperature inside the cell was monitored using a thermocouple. The ultrasound was introduced using a tapered microtip (end diameter 0.125 inch (3 mm)) driven by a Heat Systems W225 Sonicator (20 kHz, operating at 60% power).

For each reaction the copper bronze in DMF was first subjected to a period of sonication (presonication) before the addition of 2-iodonitrobenzene. In this way the copper surface could be cleaned and activated without recourse to traditional techniques (e.g. washing with EDTA) [5]. In this presonication period a noticeable brightening in the appearance of the copper powder occurred which was assumed to be due simply to the cleaning action of ultrasound. To determine if this cleaning effect would be sufficient to provide enhanced reactivity without further sonication, a fourfold excess of 'presonicated' copper was used in a normal, stirred, Ullmann coupling (Exp 1, Table 5.6). In this control reaction no more than 1.5% of the

Table 5.6 — The effect of sonication on an Ullmann coupling

Experiment	Mole ratio Cu/substrate	Reaction temperature	Agitation	Yield %[c] after 2 h
1[a]	4	62.5	Stirred	<1.5
2[a]	4	64.0	Sonicated	70.4
3[b]	4	62.0	Sonicated	81.0

[a]15-min presonication.
[b]1-h presonication.
[c]Only biaryl product and unreacted starting material were detected by GLC analysis.

product could be identified even if the presonication period was extended to 1 h. In contrast to this, 70.4% yield is obtained when sonication was continued throughout the 2 h reaction period (Exp 2), representing at least a 50-fold increase in reactivity.

It is significant that increasing the copper to an eightfold excess does not give a substantially different yield, whereas a decrease to equimolar quantities gives considerably reduced yields of about 25% in the same time. An increase in the presonication period to 1 h increases the sonochemical yield for a fourfold excess of copper (Exp 3).

Discussion

In addition to surface cleaning, ultrasonic irradiation is known to pit metallic surfaces and disperse powders into smaller particle size. This might suggest that more prolonged pre-sonication should result in even greater reactivity. The results from Exp 3 suggest that extended presonication does have an effect in increasing the yield when the reaction is irradiated throughout. Therefore the effect of ultrasound on the particle size of the copper was investigated. Samples of copper bronze were subjected to irradiation for different time periods under normal reaction conditions, but without substrate present. The v.m.d. values for the copper samples dropped from a value of 87 at time zero to 40 after 15 min and reach a limit of 25 after 30 min. These values were obtained in suspension in carbon tetrachloride using a Malvern Autosizer.

Thus ultrasonic irradiation (20 kHz directly introduced via a probe) can be shown to reduce reaction time and the amount of copper required for an Ullmann reaction. The adduct 2,2′-dinitrobiphenyl (75%) is obtained in 2 h at 62°C using only a fourfold excess of copper. This can be interpreted as a 50-fold increase in reactivity compared with non-ultrasonic conditions.

It is also noteworthy that reactions irradiated in this way do not lead to deposits of copper on the surface of the glassware, a common problem using the conventional methodology.

References

[1] P. E. Fanta, *Synthesis* (1974) 9.
[2] J. Lindley, *Tetrahedron, (1984)* **40**, 1433.
[3] M. D. Rausch, *J. Org. Chem.* (1961) **26**, 1802.
[4] J. Lindley, J. P. Lorimer and T. J. Mason, *Ultrasonics* (1986) **24**, 292.
[5] A. H. Lewin, M. J. Zovko, W. H. Rosewater and T. Cohen, *J. Chem. Soc., Chem. Commun.* (1967) 80.

5.2.5 The effect of ultrasound on the efficiency of nickel powder as a hydrogenation catalyst

from: K. S. Suslick, *School of Chemical Sciences, University of Illinois, Urbana, Illinois, USA.*

Introduction

High intensity ultrasound has been used to activate nickel powder for use as a hydrogenation catalyst. Surface studies of the metal after sonication reveal dramatic changes in morphology in that the surfaces are smoothed and, as the individual particle size is reduced, the particles begin to come together and form extended aggregates [1].

Method
All sonications were performed with a Heat Systems W375 Cell Disrupter (20 kHz) with a titanium immersion horn at acoustic intensities of 50 W cm^{-2} as described in detail elsewhere [2,3]. Irradiation in a low intensity ultrasonic cleaning bath does give hydrogenation, but at greatly reduced rates compared with the probe system [4].

Inert atmosphere work used either standard Schlenk double-manifold techniques or an inert atmosphere glove box. After freeze–vacuum–thaw degasssing of solvents it is important to thoroughly saturate with an inert gas by bubbling (preferably with argon, which increases the effectiveness of cavitational collapse). A Vacuum Atmospheres inert atmosphere glove box was equipped with an O2 monitor and maintained at less than 1 p.p.m. O_2.

All solvents for the nickel hydrogenation reactions were Gold Label purity from Aldrich Chemical Co. and used as received. All alkenes used in the reaction were obtained from commercial sources and used as received; their purity was determined to be above 99% by GLC. Ni powder (200 mesh, EM Science, Cherry Hill, NJ 08034) came packed under argon and was stored under a rigorously inert atmosphere. In a typical sonication, 4 g of metal powder were loaded into a glass sonication cell with 20 cm^3 of decane in an inert atmosphere glove box. The horn and cell assembly was then sealed, removed from the glove box and sparged with argon for 40 min prior to sonication. Sonication of the metal powder was performed at a cell temperature of 15°C for 30 or 60 min. The cell and horn assembly were then returned to the inert atmosphere box. The powder was filtered, washed with pentane and dried under vacuum at 90°C for 16–20 h. Alternatively, Schlenk air-free techniques were used and transfers made by cannula under a positive pressure of argon.

Hydrogenation reactions were carried out at 273 K under 1 atm of H_2. In a typical reaction, 1 g of Ni powder was added to a 0.5 M solution of alkene in octane. Products were analysed by gas chromatography–mass chromatography (GC–MS and capillary GC on a 30-m DB-1 column (J & W Scientific) with heptane as internal standard.

Discussion
Simple nickel powder is an extremely inactive catalyst for the hydrogenation of alkenes. Even after 2 weeks no alkane was detected when an alkene was rapidly stirred with powdered nickel, as supplied, at 273 K under H_2 gas at 1 atm pressure. In comparison (and under otherwise identical conditions) non-1-ene can be hydrogenated at up to 1.5 mM per minute if the nickel powder is first pretreated with ultrasound as described above. The effect of different pretreatment times on the initial rates of hydrogenation of non-1-ene is shown in Fig 5.12.

Ultrasonic treatment of the nickel powder for a period of 1 h gives optimum activity, which decreases with longer irradiation times. Although the results shown were obtained with non-1-ene, similar behaviour was found for a range of other alkenes.

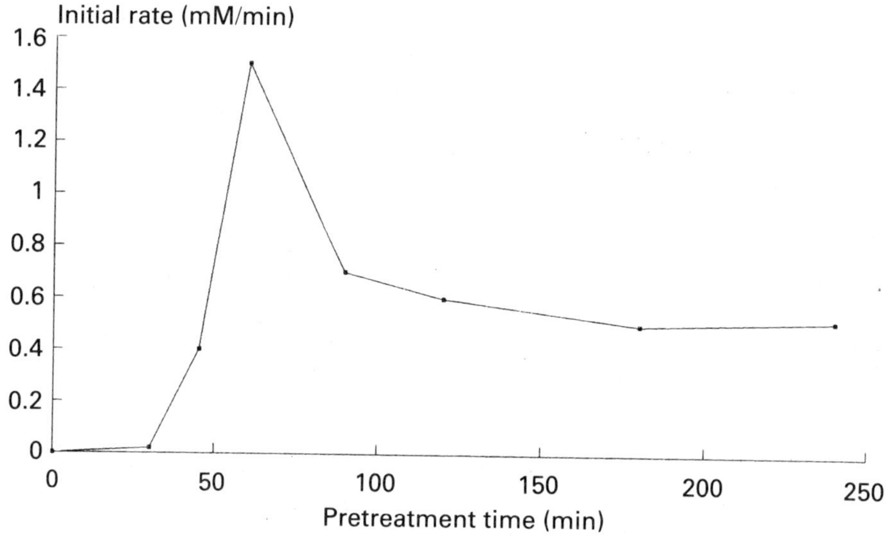

Fig. 5.12 — Effect of pretreatment time on hydrogenation rate.

No reduction of ketones or aldehydes was observed using this type of catalyst, and in no case were the irradiated powders found to be pyrophoric when exposed to air. This is somewhat different in activity from Raney nickel.

References

[1] K. S. Suslick, D. J. Casadonte and S. J. Doktycz, *Solid State Ionics* (1989) **32/33**, 444.

[2] K. S. Suslick in *High energy processes in organometallic chemistry*, Amer. Chem. Soc. Symp. Ser. 333 ed. K. S. Suslick, American Chemical Society, Washington (1987) p. 191.

[3] K. S. Suslick and E. B. Flint in *Experimental organometallic chemistry*, Amer. Chem. Soc. Symp. Ser. 357, eds A. Wayda and M. Y. Darensbourg, American Chemical Society, Washington (1987) p. 195.

[4] K. S. Suslick and D. J. Casadonte, *J. Amer. Chem. Soc.* (1987) **109**, 3459.

REACTIONS ON SOLID SUPPORTS

5.2.6 The enhancement of intercalation reactions by ultrasound

from: K. S. Suslick, *School of Chemical Sciences, University of Illinois, Urbana, Illinois, USA*.

Introduction

Ultrasound has been shown to increase dramatically the rate of intercalation of a wide range of guest molecules into layered inorganic solids [1]. Since conventional thermal intercalation can often take weeks at elevated temperatures, the benefits of using ultrasound to assist the process are obvious.

Method

In a typical experiment, a 50 cm^3 thermostatted glass vessel was charged with 0.5–1.0 g of layered solid host material and 15 cm^3 of a toluene solution of the guest material in a 2 : 1 molar excess. Under an N_2 atmosphere the solution was ultrasonically irradiated by direct immersion of a titanium horn with an acoustic intensity of *c*. 20 W cm^{-2} (Heat Systems W375, 20 kHz). Sonications were performed at 20°C and temperature control was provided with a constant temperature bath. The actual temperatures during sonication were measured by a thermocouple held in the middle of the sonication cell, free from contact with either the walls or the probe. A selection of results are shown in Table 5.7.

Table 5.7 — Comparison of thermal and sonochemical conditions for the synthesis of $[guest]_2[host]$

Conditions	Time, Temp (°C)			
$[guest]_x host$	Thermal		Sono chemical	
$[Cp_2Co]_{0.25}ZrS_2$	50 h,	20	2 h,	20
$[RNH_2]_2TaS_2$	50 h,	20	15 min,	20
$[pyridine]MoO_3$	30 days,	180	3 days,	80

Discussion

The intercalation rate enhancements induced by ultrasonic irradiation have been shown to be associated with a reduction in particle size of the host solid and with an increase in the amount of surface damage. This was investigated in more detail in the case of the hexylamine/TaS_2 system. TaS_2 in toluene was presonicated for various periods of time at 20°C and then hexylamine was added and the mixture stirred over a period of 3 h. The intercalation rate was at a maximum for a 15-min presonication time which corresponded to a reduction in particle size of the TaS_2 from 60–90 μm to 5 μm. Further sonication had little effect on particle size and caused only a small further increase in the rate of intercalation.

References

[1] K. Chatakondu, M. L. H. Green, M. E. Thompson and K. S. Suslick, *J. Chem. Soc. Chem. Commun.* (1987), 900.

5.2.7 Sonochemically assisted halogenation of aromatic hydrocarbons with alumina supported copper(II) bromide

from: The Coventry Polytechnic Sonochemistry Group.

Introduction

Copper(II) halides were first recognized as potential halogenating agents for aromatic hydrocarbons by Kochi [1]. Recently Kodomari *et al.* reported that halogenation was readily achieved by supporting the copper(II) halide on alumina [2]. In view of the interest in the use of ultrasound to enhance supported reagent reactivity we have studied a Kodomari reaction, the bromination of naphthalene to 1-bromo- and 1,4-dibromonaphthalene using $CuBr_2/Al_2O_3$ in CCl_4 (Scheme 5.22) [3].

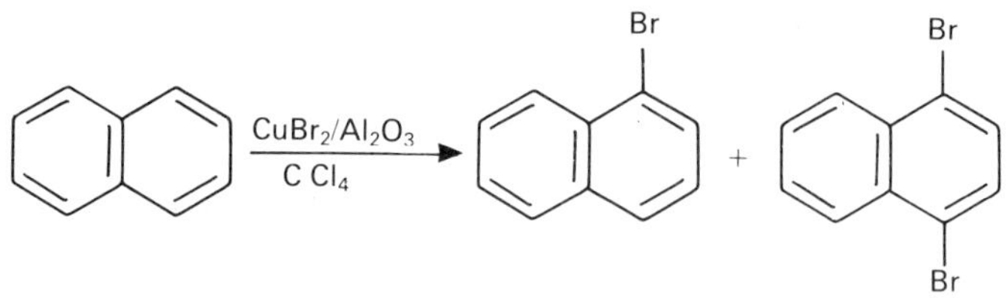

Scheme 5.22

Method

The alumina-supported copper(II) bromide was prepared conventionally by adding alumina (Aldrich 95%, activated, neutral, 40 g) to a solution of copper(II) bromide (B.D.H. general purpose, 20 g) in distilled water (60 cm^3) at room temperature. The water was then removed at reduced pressure using a rotary evaporator at 80°C and the resulting powder was dried in a vacuum oven at 100°C overnight. The product was stored in a sealed jar.

Using the same quantities of reagents, a batch of the supported material was prepared by the application of ultrasound to the suspension of alumina in aqueous copper(II) bromide for 1 h prior to drying. Insonation was provided through a Sonics and Materials VC600 Sonicator (20 kHz) operating at 42 W cm^{-2} through a horn of end diameter 0.25 inches (6 mm).

The efficiencies of these supported materials were then tested in a bromination reaction. Naphthalene (1.152 g, 9 mmol) and the supported reagent (30 g) were mechanically stirred in CCl_4 at 76°C for 4 h. Samples of the reaction mixture (1.5 cm^3) were withdrawn at 20-min intervals, filtered and the powder washed through with a further 3.5 cm^3 solvent to make a volume of 5 cm^3. These samples were analysed using a Perkin Elmer F17 gas chromatograph and an SE30 column at 175°C.

Discussion

Solid-supported reagent prepared under the influence of ultrasound was shown to be much more effective in the Kodomari reaction. Both methods gave similar product distributions, the ratio 1-bromonaphthalene : 1,4-dibromonaphthalene being 91 : 9 with normal mixing and 97 : 3 when prepared under sonication. However, ultrasonic treatment gave an 80% increase in initial rate and a reduction in time for total conversion to products from 180 min to only 60 min. Naturally, if the reaction time is extended then the proportion of dibrominated material increases as the 1-bromonaphthalene reacts further.

Electron microscopy reveals that the material prepared under sonication appears to have undergone agglomeration to give particles with a sponge-like appearance containing large cavities (Fig. 5.13).

References

[1] J. K. Kochi, *J. Am. Chem. Soc.* (1955) **77**, 5274.
[2] M. Kodomari, H. Satoh and S. Yoshitomi, *J. Org. Chem.* (1988) **53**, 203.
[3] J. Lindley, T. J. Mason and J. P. Lorimer, Presented at the First Meeting of the European Society of Sonochemistry, Autrans, France, 1990.

ORGANOMETALLIC CHEMISTRY

5.2.8 The Barbier reaction of *S*(+)2-octylbromide with cyclohexanone

from: J. C. de Souza Barboza, C. Petrier and J. L. Luche, *L.E.D.S.S., Chimie Recherche, Université Scientifique et Médical de Grenoble, France.*

Introduction

Normally the reaction of a homochiral haloalkane with a metal leads to racemic organometallic reagents [1]. A consideration of the mechanism of such reactions makes this observation relatively easy to explain by invoking the intermediacy of a rapidly inverted (or planar) free radical. Because of the possible role of ultrasonic waves in promoting electron transfers, desorbing unstable species and more generally accelerating heterogeneous reactions, the Barbier reaction was chosen as an appropriate subject for a sonochemical study [2]. Of even greater interest was the possibility that non-racemic compounds might be obtained under these conditions. The following example provides an attractive model for such studies (Scheme 5.23).

Method

The experiments were all effected with an Ultrasons Annemasse probe system (75 W, 30 kHz) equipped with a titanium horn, diameter 8 mm, and it was operated at 73% of its maximum power. The reaction vessel (see Chapter 3) proved to be of general use for sonochemical preparations [3]. It is composed of a reaction cell with a sample holder on which the lithium piece was held under the emitting face of the horn. A magnetic stirring bar was used to achieve even temperature distribution. The cell is surrounded by a cooling jacket through which the circulation of a thermostatted fluid permits good temperature control during the experiment.

Lithium (68 mg, 24 mmol, from Metallgesellschaft, containing 2% Na) was placed in the sample holder and the vessel adjusted to place this under the sonicator tip. The distance between the metal piece and the emitter was *c.* 2–3 mm. A rubber septum was used to ensure a satisfactory air seal. *S*(+)2-Bromooctane (772 mg, 4 mmol) [4] and cyclohexanone (196 mg, 2 mmol) in 25 cm^3 of dry THF were added via the side arm, and argon was introduced in the vessel. After a few minutes of

Fig. 5.13 — Effect of sonication on particle agglomeration.

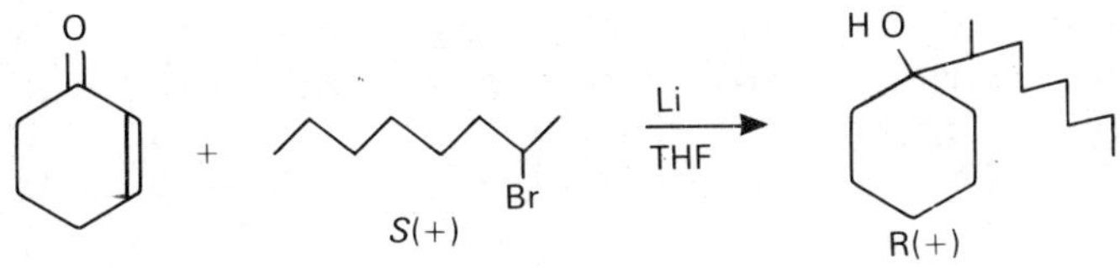

Scheme 5.23

bubbling, the flow of argon was stopped and a static atmosphere of this gas maintained through the reaction. After temperature equilibration at −50°C, as indicated by a thermocouple placed in the side arm, sonication was applied for 90 min. At the end of this period, the probe system was switched off, the vessel was removed and the excess lithium was separated. The resulting light-yellow solution was treated with 10% aqueous HCl and extracted with ether. The organic phase was washed (saturated aqueous NH_4Cl), dried over Na_2SO_4 and the solvent evaporated. The residue was purified by column chromotography, to yield an 406 mg of oil (98%) [5]; $[\alpha]_D=+3.2$ ($c=3.1$, $CDCl_3$, 20°C). After correction for the enantiomeric excess (e.e.) of the initial bromide: $[\alpha]_D=+8.2$. The alcohol was thus shown to be formed with 24% inversion of configuration [6].

Discussion

The reaction above has provided fundamental information on the mechanism of the Barbier procedure, and also on the chemical phenomena occurring during an irradiation with ultrasound. Under conventional conditions the same components, when stirred at 0°C, yield only 50% of the alcohol after 7 h. In this case the e.e. was only 6%, in comparison to the 24% obtained under sonochemical conditions. Although the stereochemical effect of sonication is significant, the stereoselectivity is still too low for synthetic applications. The effect on the chemical yield, is however, substantial, since, even at a rather low temperature, the product is obtained quantitatively.

Generalization of the procedure to other cases of Barbier reactions is straightforward, and many examples have been described in the literature [7]. The reaction vessel used allows the temperature to be easily adjusted and controlled, and the probe system permits optimization of the energy input.

References

[1] H. M. Walborsky and M. S. Aronoff, *J. Organomet. Chem.* (1973) **51**, 55.

[2] J. L. Luche and J. C. Damiano, *J. Am. Chem. Soc.* (1980) **102**, 7926; J. C. de Souza Barboza, J. L. Luche and C. Petrier, *Tetrahedron Lett.* (1987) **28**, 2013.

[3] J. C. de Souza Barboza, C. Petrier and J. L. Luche, *J. Org. Chem.* (1988) **53**, 1212.

[4] Prepared from $R(-)$octanol (Aldrich), according to: P. Roolin, *Synthetic Commun.* (1986) **16**, 611.

[5] IR: 3450, 2925, 2850, 1460, 1375, 1160, 1130, 960 cm^{-1}; nmr (80 MHz, $CHCl_3$, TMS internal reference): 1.75–1.10 (m, 22H), 0.88 (m, 6H) p.p.m.; MS m/e 212 (M+), 195, 128, 99, 81.

[6] Determined by comparison with a sample prepared from methyl *R*(−)2-methyloctanoate of known e.e. and configuration. P. A. Levene, S. A. Harris, *J. Biol. Chem.* (1936) **112**, 195.
[7] For examples, see: J. Einhorn and J. L. Luche, *Synthesis* (1989) 784.

5.2.9 The preparation of dialkylzinc reagents

from: J. L. Luche, *L.E.D.S.S., Chimie Recherche, Université Scientifique et Médical de Grenoble, France.*

Introduction

In section 5.1.1 the preparation of diaryl zinc reagents was described. These can be effected without difficulty in a cleaning bath, but for reasons which are not at all clear this experimental system was not capable of generating dialkyl compounds. These organometallic compounds were, however, quite accessible using the much more powerful horn generator so that the *n*-hexyl group can be added to cyclohexenone (Scheme 5.24) [1].

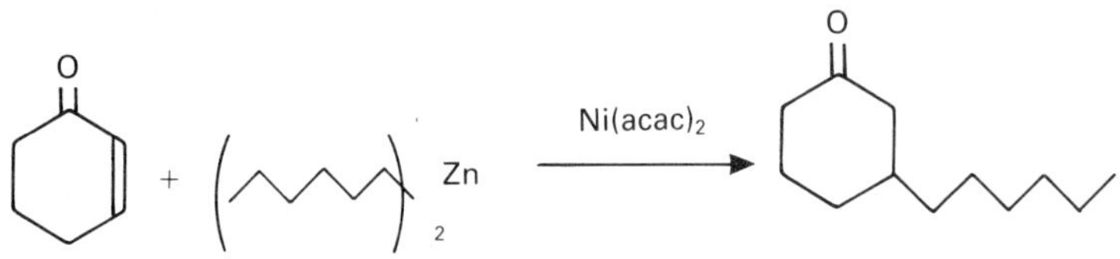

Scheme 5.24

Method

The quantities of the appropriate reagents were those employed in section 5.1.1 and the method adopted and the equipment used is described in the previous experiment (5.2.8). This methodology affords the required dialkylzinc reagent in 20–30 min as a black solution. The solution can then be withdrawn via a syringe and added to the enone substrate in the presence of nickel acetylacetonate.

Discussion

It is interesting to note that the reaction proceeds efficiently even in solvents which do not produce intense cavitation, such as diethyl ether. However, for the sake of convenience, the general method makes use of THF–toluene mixtures (1 : 10) which permit easier manipulation and drying.

Reference

[1] C. Petrier, J. L. Luche and C. Dupuy, *Tetrahedron Lett.* (1984) **25**, 3463.

5.2.10 The direct preparation of lithium diisopropylamide

from: A. De Nicola, J. Einhorn and J. L. Luche, *L.E.D.S.S., Chimie Recherche, Université Scientifique et Médical de Grenoble, France*

Introduction

In section 5.1.5 the synthetic importance of non-nucleophilic strong bases such as lithium diisopropylamide (LDA) was discussed, and their sonochemical preparations described. In order to shorten the preparation and make it economically valuable for larger-scale experiments, an alternative method of synthesis was investigated which avoided the use of a transient butyl lithium species. The principal of this procedure is shown in Scheme 5.25.

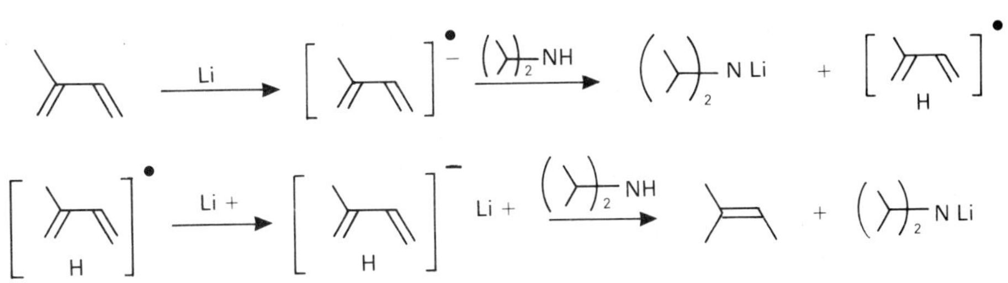

Scheme 5.25

Method

Sonications were affected using the probe technique and the apparatus described above. Lithium (0.790 g, 112 mmol, 1.5 Eq. [2% sodium, from Metallgesellshaft]) was placed in the holder just under the emitter. The vessel was charged with diisopropylamine (7.575 g, 75 mmol) in dry octane (22 cm^3) and dry THF (7.27 g, 8.2 cm^3, 1.35 Eq.). The tip of the horn was immersed in the liquid to a depth of *c*. 1 cm. The variable energy output was set to 68% of the maximum and the sonication was affected for 40 min, with the dropwise addition of 2.25 g (37.5 mmol) of isoprene via a syringe. At the end of this period, standardization of the base was performed in the usual manner and this indicated a yield of 99%.

Discussion

The direct reaction of lithium with diisopropylamine does not occur, even with sonication. An electron transfer agent is in fact necessary, and the best suited that we have found is isoprene. Styrene is used in the commercial preparation of LDA [1], but it is inconvenient in that it is transformed to ethylbenzene, which is not easily removed. It can also lead to undesired reactions in the presence of some substrates. The advantages of isoprene are essentially that it is a lighter compound (RMM=68 instead of 104 for styrene) and it is transformed to the less reactive 2-methylbutene, an easily eliminated volatile compound. In the absence of ultrasound, attempts to use this electron carrier proved to be unsatisfactory.

In this preparation, as in the case of the lithiation of 1,3-dithiane, lithium containing 2% sodium is necessary, as pure lithium reacts much more slowly. More importantly, it was noted that an optimum in the ultrasonic energy exists. At high energy (82% of the maximum) the reaction is in fact slower than at the optimum energy setting (68%). This illustrates the possible adverse effect of excessive cavitation, which disperses the energy.

Reference

[1] M. T. Reetz and W. F. Maier, *Liebigs. Ann. Chem.* (1980) 1472.

5.2.11 Preparation of transition metal carbonyl anions

from: K. S. Suslick, *School of Chemical Sciences, University of Illinois, Urbana, Illinois, USA.*

Introduction

Power ultrasound has been found to dramatically reduce the temperatures and pressures required for the preparation of early transition metal carbonyl anions from the direct reaction of the corresponding metal chlorides with carbon monoxide (Scheme 5.26) [1]. In this particular case the yield of $V(CO)_6^{2-}$ (35%) can be obtained at 4.4 atm pressure, whereas conventional methodology requires 200 atm and 160°C.

$$VCl_3(THF)_3 \xrightarrow[THF]{CO/Na} V(CO)_6^-$$

Scheme 5.26

This is representative of a range of uses for ultrasound in the amelioration of forcing conditions.

Method

All sonications were performed using a Heat Systems W-375 sonicator (20 kHz) generating acoustic intensities of 100 W cm^{-2}. The sonications were carried out in a thermostatted stainless steel cell able to operate at elevated pressure. The design is similar to that shown in Fig. 3.8.

In a typical reaction the cell was charged with 1 mmol of transition metal halide dissolved in 30 cm^3 of THF, and an excess of sodium sand was added. The mixture was flushed with CO and then maintained under CO during sonication. The system was thermostatted so that the maximum temperature reached during ultrasonic irradiation did not exceed 10°C. Sonications were carried out for 1000 min.

Discussion

This procedure works for a wide range of soluble metal halides, but the most significant are the results obtained for the synthesis of early transition metal carbonyl anions, which are normally quite difficult to prepare. The solubility of the metal halide is important, in that low yields are observed when using the halides VCl_3 or $TaCl_5$, which are only sparingly soluble in THF. Better yields result from the use of

$VCl_3(THF)_3$ in THF or $TaCl_5$ in diglyme. In all cases the yields are at least comparable to other methods requiring elevated temperatures and CO under pressure. Using this methodology approximately 50% yields of $W_2(CO)k_{10}^{2-}$, $Mo_2(CO)k_{10}^{2-}$ and $Nb(CO)k_{2-}^{6}$ are obtained from the respectivehalides.

The suggested mechanism for such carbonylations involves the direct trapping by carbon monoxide of a reactive metal species produced during the reduction of the metal halide.

Reference

[1] K. S. Suslick and R. E. Johnson, *J. Am. Chem. Soc.* (1984) **106**, 6856.

MISCELLANEOUS EXPERIMENTS USING AN ULTRASONIC PROBE

5.2.12 Ultrasonically assisted *O*-alkylation reactions of 5-hydroxychromones

from: The Coventry Polytechnic Sonochemistry Group

Introduction

The *O*-alkylation of non-hindered phenols is a facile process often achieved by reaction of the phenol with an alkyl halide, using potassium carbonate as the base, in solvents such as acetone or butan-2-one (MEK). The *O*-alkylation of 5-hydroxychromones is, however, notoriously difficult, probably as a result of hindrance to ionization caused by hydrogen bonding between the carbonyl and OH group coupled with some dispersion of the resulting phenoxide O− charge.

The substrate chosen as a model for such an *O*-alkylation was 5-hydroxy-4*H*-1-benzopyran-4-oxocarboxylic acid ethyl ester. In *N*-methylpyrrolidinone (NMP) a 28% yield of *O*-propyl product was obtained after 1.5 h at 65°C using 1-iodopropane and potassium carbonate as base (Scheme 5.27). The yield was greatly increased under sonication and the scope of the reaction was expanded by using a range of different haloalkanes [1].

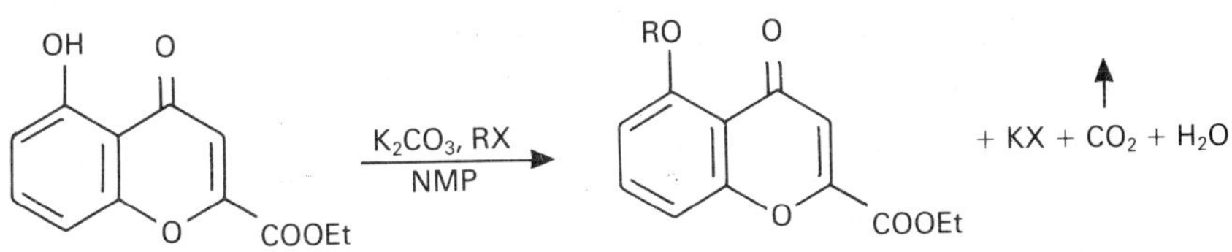

Scheme 5.27

Method:

5-Hydroxy-4-oxo-4-*H*-benzopyran-2-carboxylic acid ethyl ester (5 g, 0.021 mol) and potassium carbonate (5.95 g, 0.043 mol) were mixed with NMP (100 cm^3) in a Rosett kettle in a constant temperature bath set at 55°C. Sonication (20 kHz) was introduced via a Heat Systems W225 sonicator (operating at 38% power output, setting 4) employing a 0.5-inch (12.5 mm) titanium probe with extender fitted through a PTFE sleeve into a Rosett kettle reactor. The mixture was allowed to reach a steady temperature (65°C) before a twofold excess of the required haloalkane (0.043 mol) was added. The reaction was followed by GLC by withdrawing 1 cm^3 aliquots. These were neutralized by pouring into dilute HCl (10 cm^3) and extracting with ethyl acetate (5 cm^3). Analysis was by GLC. On completion of the reaction the mixture was poured into ice-cold 2-*N* HCl (1000 cm^3) and extracted with ethyl acetate (3×50 cm^3). The organic extract was washed with water and dried over $MgSO_4$. After solvent removal, a solid was obtained which was recrystallized from ethanol (Table 5.8). All compounds satisfied spectral, MS and elemental analysis.

Table 5.8 — Sonochemical alkylation of a 5-hydroxychromone (GLC yields, isolated yields in parenthesis)

Compound R=		Time (mins)	Yield (%) Ultrasound	Yield (%) Nonultrasound	m.p.(°C)
Propyl	(I)	90	100 (77)	28	101–103
Benzyl	(Br)	105	97 (85)	59	114–116
Methyl	(I)	30	89 (54)	28	117–119
Butyl	(I)	90	97 (84)	57	134–136
Allyl	(Br)	60	100 (73)	58	119–121

Conditions: A twofold excess of haloalkane and K_2CO_3 over hydroxychromone (5 g) in 100 cm^3 NMP at 65°C.

GLC was performed on a Shimadzu Mini 3 chromatograph using an OV17 (stainless steel) column (length 2 m, diameter 0.5 cm) using N_2 as carrier gas. The oven temperature was 260°C and injection temperature 280°C.

Discussion

A possible contributing factor to the lack of reactivity may be that the K_2CO_3 used as a base is not sufficiently soluble in the reaction mixture, thus restricting reactivity to the surface of the powder. Power ultrasound would be expected to be effective in enhancing this reaction via the reduction of the particle size of K_2CO_3 powder. This factor was investigated by first sonicating NMP containing K_2CO_3 at 65°C . The appropriate proportions of 1-iodopropane and chromone were added to the resulting very fine suspension and the reaction was run under conventional conditions. This resulted in 90% product formation in 90 min at 65°C. This reactivity is similar to that obtained under continuous ultrasonic irradiation, except that the reaction was approaching a definite limit at this stage. This fall-off suggests that the KBr produced in the reaction might mask the surface activity of the remaining K_2CO_3. Evidence for this was obtained by sonicating the reaction as it approached the limiting 90% yield. Application of ultrasonic irradiation at this stage rapidly completes the reaction.

Reference

[1] T. J. Mason, J. P. Lorimer, L. Paniwnyk, A. R. Harris and P. W. Wright, *Ultrasonics International 89, Conference Proceedings* (1989) 1275.

5.3 MISCELLANEOUS SONOCHEMISTRY EXPERIMENTS

A COMPARISON OF ULTRASONIC EQUIPMENT

5.3.1 Sonochemical enhancement of a solid–liquid phase transfer catalysed reaction: the Michael addition of diethylmalonate anion to chalcone

from: J. Berlan, H. Delmas, A. M. Wilhelm and N. Ratoarinoro, *Ecole Nationale d'Ingénieurs de Génie Chimique, 31078 Toulouse, France*

Introduction

Michael addition reactions [1] have been successfully performed on a small scale in the absence of a solvent using a phase transfer catalyst [2]. High yields are obtained under very mild conditions, without stirring. Unfortunately, however, the scale-up of these reactions is difficult, since, for larger quantities, mass transfer problems arise, due to the viscosity of the reacting mixture, which cannot be conveniently stirred. The addition of a solvent certainly lowers the viscosity of the reacting medium, but this also dramatically decreases the rate of the reaction.

Sonication offers a solution to the problem by improving the mass transfer and mixing of such reactions. This is illustrated in the case of the addition of the diethylmalonate anion to chalcone (Scheme 5.28) for which three different types of sonicator systems can be used.

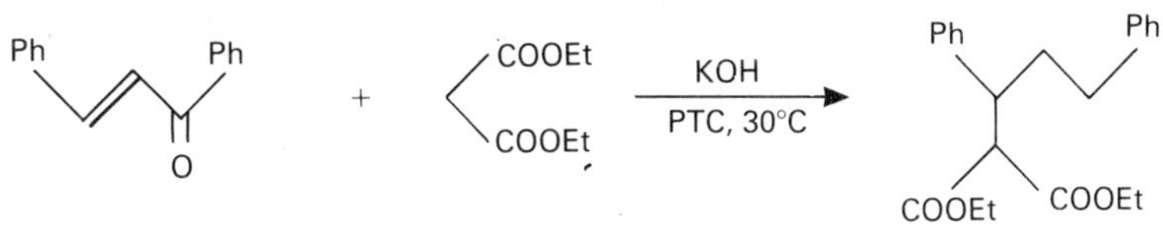

Scheme 5.28

Method 1: Sonication with a probe system

The apparatus used is shown in Fig. 5.14, incorporating a Sonics & Materials Vibracell VC600 unit (20 kHz) with a horn of end diameter 12 mm. The reaction vessel is a three-necked, flat-bottomed, jacketed reactor (internal diameter: 5 cm). Sampling of the reaction mixture is accomplished through a rubber septum with a

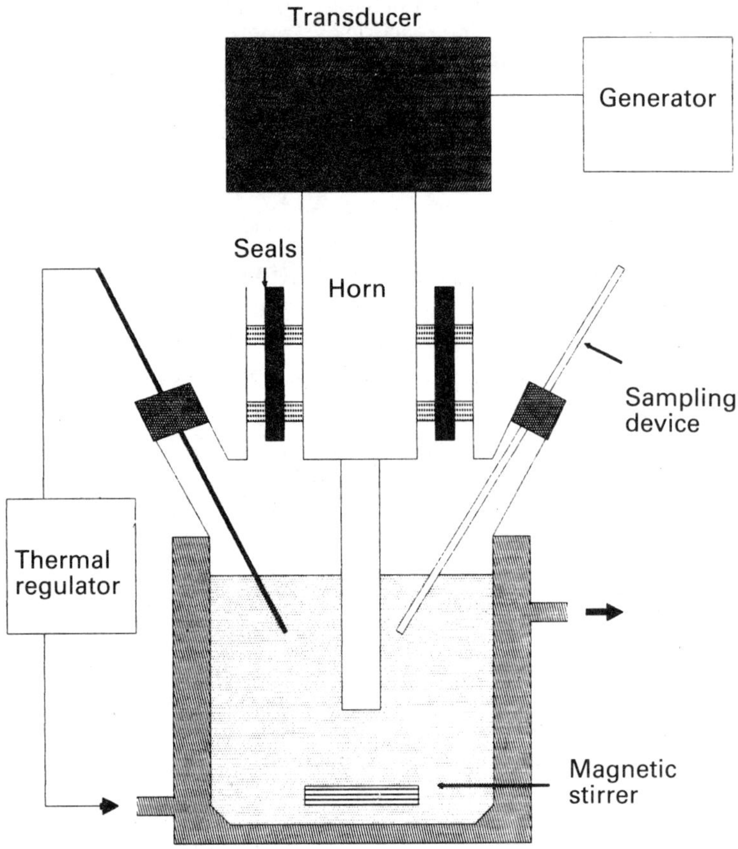

Fig. 5.14 — Temperature-controlled probe reactor.

syringe. The reaction temperature is controlled by water circulation through an external thermostatted jacket. The circulation is regulated by a platinum temperature probe dipped in the reaction itself.

The horn passes through the central neck of the reactor and is sealed into the system with a stainless steel ring fitted with viton O-seals. The distance from the horn to the bottom of the reactor is 3.7 cm.

A mixture of chalcone (20.436 g, 100 mmol), crushed potassium hydroxide (0.348 g, 6.2 mmol, particle size 200–300 μm), benzyltriethylammonium chloride (BTEA, 1.108 g, 6 mmol) and toluene (100 cm^3) were introduced into the reactor and the temperature was maintained at 30°C. Diethyl malonate (16.016 g, 100 mmol) was then added in one portion and the mixture sonicated for 2 min (power output 130 W — note 1) under magnetic stirring (at 300 r.p.m.). The depth of the probe immersion is 2.4 cm.

The solid phase was filtered off, the toluene phase washed twice with 30 cm^3 of water, and the toluene removed under vacuum. The crude product, recrystallized from 60% ethanol, was obtained in almost quantitative yield (m.p. 73°C) (see notes 2,3 and 4). No by-product was detected.

The HPLC conditions were: Waters 501 pump; mobile phase: hexane–diethyl ether 70/30 (solvents SDS, Chromasol); flow rate: 1.2 $cm^3 min^{-1}$; UV detector Milton Roy SM 3000 (λ=230 nm); column: ColoChrom 125×6 mm; packing: Spherisorb silica 5 μm.

Note 1

The power output was determined from the power output chart given in the instruction manual provided by Sonic & Materials. This power output has been checked by a calorimetric method as follows: the jacket was evacuated to ensure thermal isolation, and 100 cm^3 of water was introduced into the reactor. The water was then sonicated for 1 min under magnetic stirring. The temperature increase was 14.5°C measured with a thermocouple probe (AOIP chrome/alumel); this represents 101 W of output power. This value is in fairly good agreement with the value determined from the chart (given that thermal losses cannot be totally avoided).

Note 2

Identical results are obtained with a power output within the range 50–200 W.

Note 3

Under stirring alone (300 r.p.m.), without sonication, the yield is less than 5%. After 1 h the yield obtained is 97%.

Note 4

Without a phase transfer catalyst no reaction can be detected under stirring alone; however, sonication under these conditions affords a 95% yield within 10 min. With a high speed stirrer (15 500 r.p.m.) a 25% yield is obtained after 10 min.

Method 2: Sonication with a cup-horn system

The apparatus used was as shown in Fig. 2.8.

The same Sonic & Materials VC600 unit was used to drive the cup-horn attachment internal diameter 63 mm. The height of the liquid above the horn was 51 mm giving a liquid content of 210 cm^3. A flat-bottomed 100-cm^3 flask (diameter 5 cm, fitted with a septum for sampling) is the reaction vessel. The temperature of the reaction is measured with a platinum probe dipped into the reaction mixture itself.

A mixture of chalcone (10.418 g, 50 mmol), potassium hydroxide (0.174 g, 3.1 mmol, particle size 200–300.), BTEA (0.554 g, 3 mmol) and toluene (50 cm^3) was introduced in the reaction flask. The flask was placed in the cup, through which thermostatted water (30°C) was circulated. Diethyl malonate (8.008 g, 50 mmol) was then added in one portion, and the mixture sonicated for 2 min (power output 100 W, see note 5) without stirring. The liquid height in the flask was 3 cm.

The work-up, analysis of the crude product and the identification of the product are as described above.

Note 5

The power output was checked by a calorimetric method as follows: the cup was jacketed with glass wool to ensure partial thermal insulation and filled with 210 cm^3 of water. The ultrasound was switched on for 1 min resulting in a 6.1°C temperature increase (measured with a thermocouple probe, AOIP chromel/alumel). This corresponds to 84 W output power, in fairly good agreement with the value determined from the chart supplied with the instrument, given that the insulation was not perfect.

Method 3: Sonication with a hexagonal bath

This tailor-made stainless steel bath has been supplied by the French company Sonic Service. Power output can be varied from a maximum of 600 W to 300 W when transducers A, B and C are switched off. The frequency is 20 kHz and the total capacity of the bath is 2 l (see Fig. 4.8).

The reaction has been carried out in a 1-l cylindrical, round-bottom, glass reactor (internal diameter 10.5 cm, glass thickness 3 mm), fitted with a mechanical stirrer (see note 6). The temperature was measured with a Quick thermometer (model 8517-10) with a stainless steel probe, dipped in the reaction mixture and connected with an extension cable.

In a typical experiment the reactor was centred in the hexagonal bath and chalcone (208.26 g, 1 mol), potassium hydroxide (3.5 g, 0.06 mol), BTEA (11 g, 0.06 mol) and toluene (1 l) were mixed under stirring with a stainless steel turbine propeller at 300 r.p.m. Water at a temperature of 30°C was circulated in the bath with an external circulator maintaining a water height in the bath of 13 cm. When the temperature in the reactor attained 30°C, diethylmalonate (160.2 g, 1 mol) was added in one portion, and the mixture sonicated for 2 min at a 300-W power output (Note 7). From the HPLC analysis, pure adduct was obtained in 97% yield, based on chalcone.

In a control experiment, under the same conditions but with mechanical stirring alone, the yield was less than 5%. The work-up and analysis of the crude product were performed as above.

Note 6

This reaction can easily be carried out as a continuous flow process. Twelve litres of reacting mixture were prepared by mixing 10 mol each of chalcone and malonate with potassium hydroxide (35 g), BTEA (110 g) and toluene (10 l). The mixture was warmed to 30°C and one litre of the mixture was introduced into the reactor under stirring. The rest of the mixture was kept in a stirred thermostatted tank. The reactor was sonicated at a 300-W output power, and the reacting mixture circulated in the reactor with a double-headed Masterflex pump, at a 500 $cm^3\ min^{-1}$ flow rate during 25 min (residence time 2 min). From the HPLC analysis, pure adduct was obtained in a 98% yield, based on chalcone.

Note 7

The power output was given by Sonic Service. This power output was checked by a thermal method in the following way: the bath was filled with 1.5 l of water and power was switched on for 1 min with the liquid vigorously stirred with a turbine propeller. A 2.5°C temperature increase was measured with a thermocouple probe (AOIP chromel/alumel), giving a 261-W output power, in fairly good agreement with the value given by the manufacturers.

References

[1] J. Mathieu, J. Weil-Raynal, *Formation of C–C Bonds*, Vol 2, Georg. Thieme. Verlag, Stuttgart 152 (1975); E. D. Bergman, D. Ginsberg and P. Pappo, *Org. Reactions* (1959) **10**, 179.

[2] G. Bram, J. Sansoulet, H. Galons, Y. Bensaid, C. Combet-Farnoux and M. Mioque, *Tetrahedron Letts.* (1985) **38**, 4601.

5.3.2 The saponification of wool grease using a whistle reactor

from: A. Safdar and R. S. Davidson, *Department of Chemistry, City University, London*

Introduction

The hydrolysis of wool wax and the subsequent refining of wool wax alcohols provides an excellent example of the use of a laboratory liquid whistle reactor. Generally it is with liquid–liquid reactions involving phase transfer catalysts (PTCs) that the full benefit of ultrasonic homogenizers becomes evident. Sonication produces homogenization, i.e. very fine emulsions, which greatly increases the reactive interfacial area and allows faster reaction at lower temperatures. Davidson has reported an example of this with the ultrasonically enhanced saponification of wool waxes by aqueous sodium hydroxide using a PTC [1]. The products obtained using this technique were in higher yield than those obtained using a conventional hydrolysis and, perhaps of even more importance, the alcohols were cleaner and less coloured. This indicates that the ultrasonic method leads to less decomposition.

Wool wax esters $\rightarrow$ Wax alcohols and sodium soaps

Method

To ethanol solution (750 cm^3) wool wax C76 (93.75 g) was added, followed by aqueous sodium hydroxide (14.4 g, 2.4 M) and Aliquat (tricaprylmethylammonium chloride, 2.5% by volume). The reacting mixture was brought to 80°C and this mixture was cycled through a Lucas Dawe Minisonic homogenizer for 5 min. The reacting mixture was then transferred into a conical flask (2 l, fitted with a condenser) and maintained at 90–95°C in a water bath for 5 min.

The separation of the wool wax alcohols was achieved by extraction with petroleum ether (b.p. 60/80 in the proportion 22% petroleum ether of the total volume) and eight extractions were completed to get complete separation. The ethereal extracts were combined and the solvent was evaporated to dryness to yield the wax alcohols.

The crude extract of wax alcohols (46 g) was purified by adding it to ethanol solution (300 cm^3) and warming the solution in a water bath (60°C) for a few minutes. Ethanolic calcium chloride solution (0.5 M) was added (in excess) to the above hot solution. The mixture was well stirred, and the insoluble calcium soaps were precipitated and filtered off while the solution was still hot (55°C). Carbon dioxide gas was bubbled into the filtrate and the excess calcium ion was precipitated as carbonate. This was also filtered off and the filtrate evaporated to dryness to yield the wax alcohols (yellow in colour, m.p. 52-55°C).

After the extraction of wax alcohols, the reacting mixture was acidified (HCl) and the wax acids were extracted with petroleum ether (60/80).

Reference

[1] R. S. Davidson, A. Safdar, J. D. Spencer and D. W. Lewis, *Ultrasonics*, (1987), **25**, 35.

Index